MATHEMATICS

Foundation

- Build Strong Math foundation in no time!

- Practice with hundreds of Problems!

- Be Pre-algebra Ready!

Sareedo Pre-University Books

Contents

Welcome Students!

- You did the right choice in deciding to work with this guide. Your time is very valuable and I don't want you to waste it in long exercises and tedious explanations that do not add any meaningful understanding to the concepts and skills required to master the foundations of math.

- Instead, this guide is just straight to the point! Each and almost every page starts with quick and short example or explanation immediately followed by the skill practice. Work on these exercises and check the selected answers (usually odd numbers) at the back of the book!

- That is it. In no time, you master the math concepts and be ready to cruise algebra at high speed!

All the best!

UNIT 1: THE FOUNDATION

1.1) Review of Basic Operations:

1.2) Properties of Addition

1.3) Properties of Multiplication

1.4) Order of Operations

1.5) Review of Primes and Composite?

1.6) Prime Factorization

1.7) Factoring

1.8) Review of Perfect Squares

1.9) Square Roots

A tiger can carry about 550kg, twice its own body weight because the tiger's legs are so powerful. What do you think it will take to build strong math legs or foundation?

1.1) REVIEW OF BASIC OPERATIONS:

a) Review of Additions

1)

```
  3 1 2          2 4 5              7 5          4 6 9
+   6 0        +   5 7          + 6 8 3        +     9
```

2)

```
  1 5 2          5 2 4          6 1 7          5 4 9
+ 9 6 7        + 2 5 8        + 6 8 6        + 7 9 8
```

3)

```
  7 1 2          2 7 4          1 7 6          4 9 3
+ 5 6 5        + 2 5 6        + 6 8 8        + 7 0 9
```

4)

```
  5 7 1 2        2 0 1 4        8 7 1 6        4 9 3 6
+ 5 6 5 9      + 2 5 6 7      + 6 3 2 4      + 7 0 9 5
```

5)

```
  2 5 9 9        2 4 9 8        2 5 6 7        5 2 5 8
  7 6 2 5        4 6 0 5        5 6 2 5        6 6 2 5
+     4 6      +   7 3 7      + 7 7 4 6      + 4 5 6 6
```

b) Review of Subtractions

1)

$$
\begin{array}{r}
3\ 1\ 2 \\
-\quad 2 \\
\hline
\end{array}
\qquad
\begin{array}{r}
2\ 4\ 5 \\
-\quad 3\ 3 \\
\hline
\end{array}
\qquad
\begin{array}{r}
8\ 7\ 5 \\
-\quad 8\ 3 \\
\hline
\end{array}
\qquad
\begin{array}{r}
4\ 6\ 6 \\
-\quad 8 \\
\hline
\end{array}
$$

2)

$$
\begin{array}{r}
5\ 0\ 2 \\
-\quad 6\ 7 \\
\hline
\end{array}
\qquad
\begin{array}{r}
5\ 2\ 4 \\
-\quad 5\ 8 \\
\hline
\end{array}
\qquad
\begin{array}{r}
6\ 1\ 7 \\
-\quad 8\ 6 \\
\hline
\end{array}
\qquad
\begin{array}{r}
5\ 0\ 9 \\
-7\ 9\ 8 \\
\hline
\end{array}
$$

3)

$$
\begin{array}{r}
7\ 1\ 2 \\
-5\ 6\ 5 \\
\hline
\end{array}
\qquad
\begin{array}{r}
8\ 7\ 4 \\
-2\ 5\ 6 \\
\hline
\end{array}
\qquad
\begin{array}{r}
9\ 7\ 6 \\
-6\ 8\ 8 \\
\hline
\end{array}
\qquad
\begin{array}{r}
4\ 9\ 3 \\
-1\ 0\ 9 \\
\hline
\end{array}
$$

4)

$$
\begin{array}{r}
5\ 7\ 0\ 2 \\
-5\ 6\ 5\ 9 \\
\hline
\end{array}
\qquad
\begin{array}{r}
7\ 0\ 0\ 4 \\
-2\ 5\ 6\ 7 \\
\hline
\end{array}
\qquad
\begin{array}{r}
8\ 7\ 1\ 6 \\
-6\ 3\ 2\ 4 \\
\hline
\end{array}
\qquad
\begin{array}{r}
9\ 0\ 0\ 6 \\
-7\ 0\ 9\ 5 \\
\hline
\end{array}
$$

5)

$$
\begin{array}{r}
5\ 7\ 1\ 2 \\
-\quad 9\ 9\ 9 \\
\hline
\end{array}
\qquad
\begin{array}{r}
4\ 0\ 1\ 4 \\
-2\ 5\ 6\ 7 \\
\hline
\end{array}
\qquad
\begin{array}{r}
8\ 0\ 0\ 0 \\
-6\ 3\ 2\ 4 \\
\hline
\end{array}
\qquad
\begin{array}{r}
8\ 9\ 3\ 6 \\
-7\ 0\ 9\ 5 \\
\hline
\end{array}
$$

c) Review of Multiplications

1)

3 4 2	2 4 5	9 7 5	4 6 6
× 2	× 5	× 3	× 8

2)

5 0 2	5 2 4	6 1 7	5 0 9
× 6 7	× 5 8	× 8 6	× 9 8

3)

7 1 2	8 7 4	9 7 6	4 9 3
×5 6 5	×2 5 6	×6 8 8	×1 0 9

4)

5 7 0 2	7 0 0 4	8 7 1 6	9 0 0 6
× 5 9	× 5 6 7	×6 3 2 4	×7 0 9 5

d) Review of Divisions

1) $8\overline{)128}$ $7\overline{)917}$ $9\overline{)855}$ $8\overline{)432}$

2) $6\overline{)6918}$ $2\overline{)618}$ $3\overline{)8496}$ $8\overline{)8640}$

3) $14\overline{)824}$ $22\overline{)894}$ $13\overline{)882}$ $83\overline{)648}$

4) $62\overline{)6918}$ $24\overline{)7618}$ $33\overline{)8496}$ $81\overline{)8640}$

1.2) PROPERTIES OF ADDITION

Identity Property of Addition: adding zero to a number changes nothing	$0 + 7 = 7 \ or \ x + 0 = x$

Commutative Property of Addition: changing the order of what is being added doesn't change the result.	$5 + 7 = 7 + 5 \ or \ a + b = b + a$

Associative Property Addition changing the grouping of what is being added doesn't change the result.	$(5 + 3) + 7 = 5 + (7 + 3)$ $(a + b) + c = a + (b + c)$

How to use the properties of Addition: knowing the rules helps us to do mental math quickly. See examples

1) Find 7 +19 +13=? (It is easy to add 7 to 13 which is 20; then 20+19 is easy =39)

2) Find 44 +53 +6 =? (It is easy to add 44 to 6 which 50; then 50+53 is easy =103)

Practice:

Use mental math to find the sum:

1) 14 + 23 + 7

2) 29 + 17+ 12

3) 15 + 17 + 25

4) 49 + 14 +16

5) (162 + 235) + 38

6) 15 + (63 + 85)

Use mental math to find the sum and difference

7) 17 – 23 + 13

8) 10 – 23 + 43

9) 29 -17- 2

10) 49 + 14 - 19

11) (162 - 19) + 20

12) 8.5 + (6.3 – 8.5)

1.3) PROPERTIES OF MULTIPLICATION

Identity Property of Multiplication: Multiplying with 1 changes nothing	$1 \times 7 = 7 \ or \ x \times 1 = x$

Commutative Property of Multiplication: changing the order of what is being multiplied doesn't change the result.	$5 \times 7 = 7 \times 5 \ or \ a \times b = b \times a$

Associative Property Multiplication: changing the grouping of what is being multiplied doesn't change the result.	$(2 \times 3) \times 5 = (2 \times 5) \times 3$ $(a \times b) \times c = a \times (b \times c)$
Zero Property Multiplication: Multiplying any number with zero changes the results into zero.	$2 \times 0 = 0 \quad a \times 0 = 0$

How to use the properties of Addition: knowing the rules helps us to do mental math quickly. See examples

1) Find 4 × 7 × 25=? (It is easy to multiply 4 to 25 which is 100; then × 7= 700

2) Find 20 × 17 × 5 =? (It is easy to multiply 20 to 5 which is 100; then × 17= 1700

Practice:

Use mental math to find the product:

1) 5 × 23 × 2

2) 25 × 17× 4

3) 8 × 16 × 5

4) 49 + 14 +16

5) (4 × 35) × 5

6) 20 × (67 × 5)

7) 6 ×(9×5)

8) 6×(50×7)

1.4) ORDER OF OPERATIONS

Example: simplify $12 \div 4 \times (7 - 2) + 6$

1. Work inside grouping symbols which are: parentheses ( ), brackets [ ], and braces { } 2. Multiply and divide in order from left to right 3. Add and Subtract in order from left to right	$= 12 \div 4 \times 5 + 6$ $= 3 \times 5 + 6$ $= 15 + 6$ $= 21$

Simplify Using the Order of Operations:

1) $16 - 4 + 25 - 17$

2) $9 \div 3 \times 2$

3) $8 \div (4 \div 2^2) + 6$

4) $54 \div (32 - 31 + 2)$

5) $12 \div 6 \div 2$

6) $50 \div 5 \times 10 + 3$

7) $(7 - 5)^2 + 2$

8) $2^4 \div 4 - 1 + 2$

9) $16 \div 2(6 - 2)$

10) $4 \times 2^3 - 2(6 - 2) + 3^2$

Order of Operations:: Extra Practice

1) $(5-3)^3 \times 2 - 5$

2) $3 \times [2 \times (6-3)^2] + 6(9 \div 3)^3$

3) $8 - 2[11 - (8 \div 2^2)] + 6$

4) 4) $50 \div 5 \times 10 + 3$

5) $10 \times 2^3 \div 4 - 21$

6) 6) $14 \div 7 \times (4 \div 2) \times 3$

7) $[5 \times (6-3)^2] + 6(6 \div 3)^3$

8) $7 + 3 \times 2^4 \div 5 + \sqrt{81}$

9) $392 \div (-4-3)^2 - 2(16 \div 8)^2$

10) $18 \div 3 \times 3 \div 2 \times 3 \sqrt{16}$

1.5) REVIEW OF PRIMES AND COMPOSITE?

A **Prime number** is a whole number that can be divided evenly by 1 and itself only. For example 7 can be divided evenly by 1 or 7 only.

But, **8** can divide evenly by1, 2, 4 and 8. So, it is not prime number. It is a **composite**.

Determine whether the following numbers are primes or composite:

1) **7**	Prime		11) **18**	Composite		21) **79**	Prime
2) **9**	Composite		12) **43**			22) **12**	Composite
3) **10**			13) **59**			23) **77**	
4) **13**			14) **61**			24) **99**	
5) **1**			15) **92**			25) **11**	
6) **17**			16) **88**			26) **73**	
7) **25**			17) **39**			27) **83**	
8) **21**			18) **75**			28) **97**	
9) **31**			19) **45**			29) **29**	
10) **23**			20) **90**			30) **49**	

1.6) PRIME FACTORIZATION

Example: Find the prime factorization of: (a) 84;   (b) 138
Divide the number continuously until you have no other number to divide to it except to itself and 1!

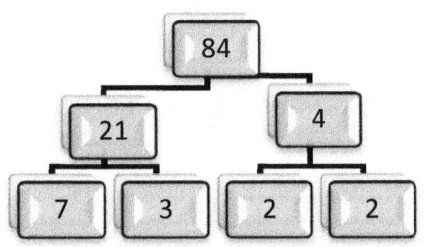
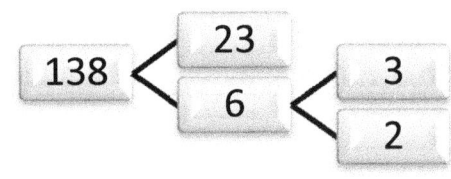

(a) The answer: $2^2 \times 3 \times 7$

(b) The answer: $2 \times 3 \times 23$

1) 16

2) 48

3) 56

4) 63

5) 39

6) 112

7) 100

8) 96

9) 75

10) 325

11) 128

12) 396

1.7) FACTORING

To factor a number means to break the number as a product of other numbers. The way to find the factor is to find which numbers that can divide evenly the number. Find the factor pairs and list!

Example 1: Find the factors of 40	Factor pairs
40 ÷1 = 40	1, 40
40÷2 = 20	2, 20
40 ÷ 4 = 10	4, 10
40÷ 5 = 8	5, 8

So, the factors are:
1, 2, 4, 5, 8, 10,20, and 40
All these numbers divide 40 evenly

Example 2: Find the factors of 42	Factor pairs
42 ÷ 1 = 42	1, 42
42 ÷2 = 21	2, 21
42 ÷3 =14	3, 14
42 ÷ 6 = 7	6, 7

So, the factors are:
1, 2, 3, 6, 7, 14, 21, and 42.
All these numbers divide 42 evenly.

1) 16

2) 48

3) 56

4) 63

5) 84

6) 25

7) 136

8) 105

9) 35

10) 117

11) 100

12) 75

1.8) REVIEW OF PERFECT SQUARES

If you know these squares on the top of your head, you will have easy time with the exponents and square roots!

Exercise 1: Simplify the following squares;

$0^2 =$	$1^2 =$	$2^2 =$	$3^2 =$
$4^2 =$	$5^2 =$	$6^2 =$	$7^2 =$
$8^2 =$	$9^2 =$	$10^2 =$	$11^2 =$
$12^2 =$	$13^2 =$	$14^2 =$	$15^2 =$
$16^2 =$	$17^2 =$	$18^2 =$	$19^2 =$
$20^2 =$	$21^2 =$	$22^2 =$	$23^2 =$
$24^2 =$	$25^2 =$	$26^2 =$	$27^2 =$

To get stronger in mental math, memorize these perfect squares.

1.9) SQUARE ROOTS

Square roots are the inverse of the squares

Since $5^2 = 25$, the square root of 25 is 5

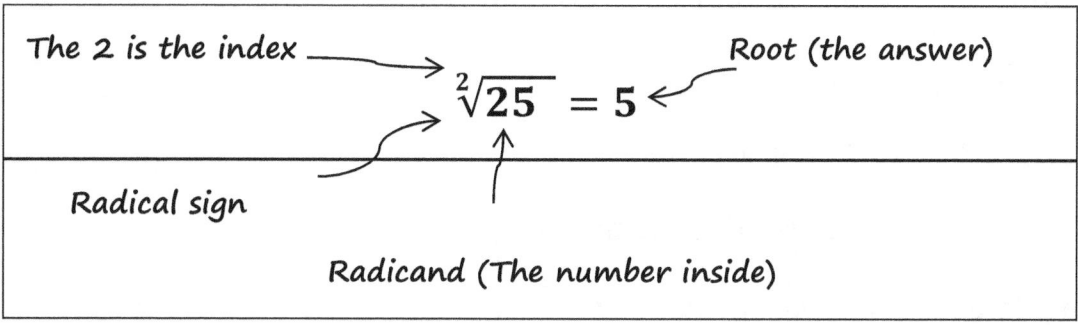

The 2 is the index

Root (the answer)

$\sqrt[2]{25} = 5$

Radical sign

Radicand (The number inside)

Note: The square root is so common that we usually omit the indices or the 2 and write simply it as: $\sqrt{25}$

Find the square roots of the following:

$\sqrt{16}$	$\sqrt{49}$	$\sqrt{1}$	$\sqrt{36}$
$\sqrt{9}$	$\sqrt{4}$	$\sqrt{0}$	$\sqrt{64}$
$\sqrt{81}$	$\sqrt{100}$	$\sqrt{169}$	$\sqrt{121}$
$\sqrt{400}$	$\sqrt{256}$	$\sqrt{144}$	$\sqrt{196}$
$\sqrt{225}$	$\sqrt{361}$	$\sqrt{289}$	$\sqrt{324}$

To get stronger in mental math, memorize these perfect square roots.

UNIT 2: FRACTIONS REVIEW

2.1) Find Greatest Common Factors (GCF):

2.2) Find the Least Common Multiple (LCM)

2.3) Review Of Fractions: Improper & Mixed

2.4) Simplify Fractions

2.5) Add/Subtract Fractions

2.6) Multiply Fractions & Mixed Numbers

2.7) Divide Fractions/Mixed Numbers

Let's Face It. Fractions are not that strong! Smash the fractions .The rest of math will then be just softer!

2.1) FIND GREATEST COMMON FACTORS (GCF):

A) Find the GCF using Factor Lists Method

The name says all! To find the greatest factor (GCF) of two numbers,

(1) First, list the factors of each number

(2) Second, choose the greatest common number!

Example: Find the greatest common factor of 16 and 28

List the factors of each: {28 => 1,2,3,**4**,7,28   {16 => 1,2,3,**4**,8,16

See the common factors are 1, 2, 3, and 4. But we choose the greatest which is **4**.

		List the Factors here					The GCDF is
1)	**(6,8)**	6:	1	**2**	3	6	2
		8:	1	2	4	8	
2)	**(12,16)**	12:					
		16					
3)	**(7,28)**	7:					
		28:					
4)	**(8,14)**	8:					
		14:					
5)	**(11, 22)**	11:					
		22:					
6)	**(16, 48)**	16:					
		48:					
7)	**(9,16)**	9:					
		16:					
8)	**(12,24)**	12:					
		24:					
9)	**(10, 20)**	10:					
		20:					

B) Find GCF: Use Division Ladder

	EXAMPLE A) 4 *and* 10		EXAMPLE B) 9 *and* 12	
✓ Continue to divide (the 4, and 10) or the (9 and 12) into numbers such as 2, 3, 7 etc. Until you can't find common number between numbers! ✓ Multiply the first raw numbers		4 / 10 ÷2 | 2 / 5		9 / 12 ÷3 | 3 / 4
	Since the 2 and 5 have no common number: The GCF = 2		Since the 3 and 4 have no common number: GCF = 3	

1. $(7, 21)$

2. $(12, 10)$

3. $(9, 18)$

4. $(15, 5)$

5. $(14, 8)$

6. $(20, 15)$

7. $(24, 16)$

8. $(12, 18)$

9. $(24, 18)$

10. $(6, 18)$

11. $(10, 30)$

12. $(18, 30)$

2.2) FIND THE LEAST COMMON MULTIPLE (LCM):

A) Find the LCM using Factor Lists Method

The name says all! To find the lowest common multiple, just continuously multiply 1, 2, 3 etc. to each number till you get a common multiple:

Example: Find the lowest common multiple of 6 and 8
List the multiples: {6 => 6, 12, **24**, 30   {8 => 8, 16, **24**   LCM = 24

See the smallest (least) common multiple is **24**.

		List the multiples here	The LCM is
1)	**(9,15)**	9: 18 27 36 **45** 15: 30 **45**	**45**
2)	**(12,16)**	12: 16	
3)	**(7,28)**	7: 28:	
4)	**(8,14)**	8: 14:	
5)	**(11, 22)**	11: 22:	
6)	**(16, 48)**	16: 48:	
7)	**(9,16)**	9: 16:	
8)	**(12,24)**	12: 24:	
9)	**(10, 20)**	10: 20:	

B) Find LCM: Use Division Ladder

	EXAMPLE A) 4 and 10		EXAMPLE B) 9 and 12	
✓ Continue to divide the 4, and 10 or the 9 and 12 into numbers such as 2, 3, 5 etc. Until all bottom numbers turn into ones!				
	4	**10**	**9**	**12**
	÷2	5	÷3	4
	2		3	
✓ If the number is not divisible just copy it again until you divide it.	÷2	5	÷3	4
	1		1	
✓ Multiply the first raw numbers.	÷5	1	÷4	1
	LCM = 2 × 2 × 5 = 20		LCM = 3 × 3 × 4 = 36	

1. (7, 14)

2. (6, 10)

3. (9,18)

4. (15, 5)

5. (14, 8)

6. (20,10)

7. (20, 16)

8. (12, 18)

9. (24,18)

10. (6,18)

11. (10, 30)

12. (18, 30)

2.3) REVIEW OF FRACTIONS: IMPROPER & MIXED

1) Change the improper fraction into mixed numbers:	$\dfrac{13}{3}$	$3\overline{)13}$ with 4 above, $\dfrac{12}{1}$ below	$\dfrac{13}{3} = 4\dfrac{1}{3}$

1) $\dfrac{7}{4}$

2) $\dfrac{51}{2}$

3) $\dfrac{17}{3}$

4) $\dfrac{49}{8}$

5) $\dfrac{27}{8}$

6) $\dfrac{13}{3}$

7) $\dfrac{17}{2}$

8) $\dfrac{51}{5}$

9) $\dfrac{75}{6}$

10) $\dfrac{93}{8}$

11) $\dfrac{23}{6}$

12) $\dfrac{69}{4}$

13) $\dfrac{87}{7}$

14) $\dfrac{96}{9}$

15) $\dfrac{33}{5}$

2) Change the mixed numbers into improper fraction:	$7\dfrac{3}{4}$	$4 \times 7 + 3 = 31$	$7\dfrac{3}{4} = \dfrac{31}{4}$

16) $3\dfrac{1}{4}$

17) $1\dfrac{3}{13}$

18) $10\dfrac{1}{2}$

19) $6\dfrac{1}{2}$

20) $4\dfrac{4}{5}$

21) $11\dfrac{3}{4}$

22) $8\dfrac{2}{3}$

23) $7\dfrac{3}{5}$

24) $12\dfrac{2}{7}$

25) $5\dfrac{4}{5}$

26) $8\dfrac{4}{9}$

27) $13\dfrac{3}{5}$

2.4) SIMPLIFY FRACTIONS

3) **Simplify Fractions:**	$\dfrac{16}{24}$	Divide **both** the denominator and numerator by the greatest common factor(GCF):	$\dfrac{16 \div 8}{24 \div 8} = \dfrac{2}{3}$

1) $\dfrac{2}{8}$

2) $\dfrac{3}{9}$

3) $\dfrac{5}{10}$

4) $\dfrac{4}{12}$

5) $\dfrac{5}{15}$

6) $\dfrac{6}{16}$

7) $\dfrac{7}{21}$

8) $\dfrac{8}{24}$

9) $\dfrac{9}{36}$

10) $\dfrac{10}{20}$

11) $\dfrac{25}{40}$

12) $\dfrac{35}{50}$

13) $\dfrac{40}{50}$

14) $\dfrac{65}{15}$

15) $\dfrac{33}{44}$

4) **Simplify Mixed Numbers**	$8\dfrac{6}{24}$	Divide **both** the denominator and numerator of the fraction part by the greatest common factor(GCF):	$8\dfrac{6 \div 6}{24 \div 6} = 8\dfrac{1}{4}$

16) $3\dfrac{2}{4}$

17) $7\dfrac{6}{10}$

18) $4\dfrac{2}{8}$

19) $5\dfrac{6}{9}$

20) $1\dfrac{3}{12}$

21) $4\dfrac{4}{20}$

22) $7\dfrac{5}{25}$

23) $8\dfrac{4}{10}$

24) $10\dfrac{8}{22}$

25) $11\dfrac{12}{16}$

26) $12\dfrac{14}{28}$

27) $13\dfrac{30}{50}$

2.5) ADD/SUBTRACT FRACTIONS:

A) Like Denominator

	Example 1	Example 2
1. **Add/subtract** the whole Numbers. 2. **Add/subtract** the numerators. 3. **Keep** the denominator unchanged. 4. **Simplify/rename** as needed.	$\frac{3}{8} + \frac{1}{8}$ $= \frac{3+1}{8} = \frac{4}{8}$ $\frac{4 \div 4}{8 \div 4} = \frac{1}{2}$	$3\frac{3}{10} + 1\frac{2}{10}$ $= 4\frac{5}{10}$ $4\frac{5 \div 5}{10 \div 5} = 4\frac{1}{2}$

1) $\frac{1}{5} + \frac{2}{5}$

2) $\frac{2}{7} + \frac{3}{7}$

3) $\frac{3}{8} + \frac{1}{8}$

4) $\frac{5}{9} - \frac{4}{9}$

5) $\frac{5}{12} + \frac{1}{12}$

6) $6\frac{3}{5} - 3\frac{1}{5}$

7) $1\frac{1}{5} + 3\frac{2}{5}$

8) $5\frac{1}{4} + 2\frac{1}{4}$

9) $2\frac{2}{10} + 7\frac{1}{10}$

10) $6\frac{2}{5} - 5\frac{1}{5}$

11) $\frac{9}{21} - \frac{2}{21}$

12) $\frac{3}{11} + \frac{5}{11}$

13) $\frac{3}{10} + \frac{6}{10}$

14) $\frac{4}{15} + \frac{3}{15}$

B) Add/Subtract Unlike Denominators

Examples: $\frac{1}{4} + \frac{1}{3}$ $2\frac{1}{2} + 1\frac{1}{5}$

| 1. Find the LCM.
2. Make the denominators equal to the LCM by multiplying right numbers.
3. Add/ Subtract and Simplify | The LCM is **12**

$= \dfrac{1 \times ③}{4 \times ③} + \dfrac{1 \times ④}{3 \times ④}$

$\dfrac{3}{12} + \dfrac{4}{12} = \dfrac{7}{12}$ | The LCM is **10**

$2\dfrac{1 \times ⑤}{2 \times ⑤} + 1\dfrac{1 \times ②}{5 \times ②}$

$2\dfrac{5}{10} + 1\dfrac{2}{10} = 3\dfrac{7}{10}$ |

1) $\frac{1}{2} + \frac{7}{8}$

2) $\frac{3}{4} - \frac{1}{5}$

3) $\frac{2}{5} + \frac{1}{4}$

4) $\frac{2}{4} - \frac{2}{8}$

5) $\frac{1}{4} - \frac{1}{12}$

6) $\frac{2}{3} + \frac{2}{9}$

7) $\frac{1}{4} + \frac{2}{3}$

8) $\frac{5}{8} - \frac{1}{4}$

9) $\frac{4}{15} + \frac{7}{10}$

10) $2\frac{1}{2} + 1\frac{1}{4}$

11) $5\frac{2}{4} - 2\frac{1}{12}$

12) $4\frac{1}{5} + 3\frac{1}{3}$

13) $7\frac{2}{3} - 2\frac{2}{9}$

14) $1\frac{5}{9} + 1\frac{2}{5}$

15) $7\frac{3}{5} - \frac{2}{7}$

2.6) MULTIPLY FRACTIONS & MIXED NUMBERS

	Example 1	Example 2
1. **Change** any mixed number into improper fractions. 2. **Multiply** the numerators and denominators. (Use cancelations if possible). 3. **Simplify**	$\frac{3}{8} \times \frac{2}{5}$ $= \frac{6}{40} = \frac{3}{20}$	$2\frac{1}{10} \times 1\frac{3}{7}$ $= \frac{21^3}{10_1} \times \frac{10^1}{7^1} = \frac{3}{1} = 3$

1) $\frac{1}{5} \times \frac{2}{5}$

2) $\frac{2}{7} \times \frac{3}{4}$

3) $\frac{3}{5} \times \frac{5}{8}$

4) $\frac{5}{6} \times \frac{4}{15}$

5) $\frac{2}{21} \times \frac{7}{10}$

6) $\frac{3}{10} \times \frac{4}{3}$

7) $\frac{3}{14} \times \frac{7}{18}$

8) $\frac{5}{6} \times \frac{3}{20}$

9) $\frac{3}{8} \times \frac{7}{4}$

10) $1\frac{1}{5} \times 1\frac{1}{4}$

11) $5\frac{1}{5} \times 2\frac{1}{12}$

12) $4\frac{1}{5} \times 3\frac{1}{3}$

13) $2\frac{2}{9} \times 2\frac{7}{10}$

14) $1\frac{5}{9} \times 1\frac{2}{7}$

15) $3\frac{1}{2} \times 2\frac{4}{7}$

2.7) DIVIDE FRACTIONS/MIXED NUMBERS

Example: Simplify

$$1\frac{1}{5} \div 2\frac{2}{5}$$

1. Rename any mixed numbers 2. Change the division sign into multiplication sign 3. Flip the divisor (the second fraction) & simplify	$\dfrac{6}{5} \div \dfrac{12}{5}$ $\dfrac{6}{5} \times \dfrac{5}{12} = \dfrac{30 \div 30}{60 \div 30} = \dfrac{1}{2}$		**When it is possible,** cross cancelation is easier $\dfrac{6^1}{5_1} \times \dfrac{5^1}{12_2} = \dfrac{1}{2}$

1) $\dfrac{1}{5} \div \dfrac{5}{2}$

2) $\dfrac{4}{5} \div \dfrac{8}{5}$

3) $\dfrac{2}{7} \div \dfrac{3}{14}$

4) $\dfrac{5}{9} \div \dfrac{5}{18}$

5) $\dfrac{1}{2} \div \dfrac{5}{2}$

6) $\dfrac{3}{4} \div \dfrac{9}{4}$

7) $\dfrac{9}{8} \div \dfrac{9}{8}$

8) $\dfrac{7}{25} \div \dfrac{2}{5}$

9) $\dfrac{5}{9} \div \dfrac{3}{7}$

10) $\dfrac{9}{11} \div \dfrac{3}{22}$

11) $5\dfrac{2}{3} \div 4\dfrac{6}{7}$

12) $6\dfrac{3}{4} \div 3\dfrac{3}{4}$

13) $4\dfrac{1}{5} \div 1\dfrac{1}{6}$

14) $9\dfrac{1}{8} \div 2\dfrac{5}{6}$

15) $1\dfrac{1}{5} \div 3\dfrac{1}{4}$

UNIT 3: THE DECIMALS

Just think about it:

For thousands of years people divided their days in 12 parts like 12 hours. So each part is about 1/12 or 0.0833333 of the day. Wouldn't it be better if we just divide the day into 10 equal parts and thus make it 10 hours a day? So, each part is just 1/10 or 0.1? Can you think of other base 12 situations?

3.1) How to Add or Subtract Decimals

Examples:

a) 3.48 + 0.15 + 27.03   **b)** 1.4 + 0.0 05 + 22.03

Step by Step   Examples

Step by Step		
Step1: Align the decimal points.	a)	b)
	3.4 8 0	1.4 00
Step 2: Add or subtract as whole numbers.	0.1 5 0	0.0 05
Step 3: Put the decimal straight down into the answer:	+2 7.0 3 1	+22.0 30
	3 0 . 6 61	23.435

<u>Note:</u> **We have added the zeros marked as 0 to help us in the alignment.**

Check point: Add or Subtract as shown above:

1. 3.4 + 2.1

2. 11 + 0.15 + 3.5

3. 3.49 + 0.15

4. 1 + 0.05 + 3.6

5. 98.04 − 1.5

6. 1.1 + 0.015 + 4.5

7. 11 − 0.15

8. 11 + 0.15 + 3.5

3.2) ADD OR SUBTRACT DECIMALS

Add the Decimals: Show your work

1. $8.4 + 6.1$

2. $10 + 0.12 + 2.5$

3. $5.49 + 0.17$

4. $7 + 0.04 + 3.6$

5. $98.04 + 1.5$

6. $1.1 + 0.015 + 0.45$

7. $11 + 0.15$

8. $15 + 0.8 + 6.5$

Subtract the Decimals: Show your work

9. $6.4 - 2.1$

10. $10.15 - 3.5$

11. $3.89 - 0.15$

12. $1 - 0.05$

13. $98.04 - 1.5$

14. $11.4 - 4.5$

15. $13 - 0.09$

16. $26 - 0.15$

3.3) MULTIPLICATION OF DECIMALS

Examples: 3.8×0.21

Step 1: Ignore the decimals & Multiply	3.8	One decimal place
	0.21	+ Two decimal places
2) Insert the decimals *(Use the total number of decimal places)*	38 760 .798	= Three Decimal places

Note: since there is no number before the decimal *you can put zero before it so that we could see the decimal better:* **0.798**!

1)
$$
\begin{array}{r} 7\,4.2 \\ \times \quad\ 2 \\ \hline \end{array}
\qquad
\begin{array}{r} 0.\,4\,8 \\ \times \quad\ 5 \\ \hline \end{array}
\qquad
\begin{array}{r} 9\,.6\,1 \\ \times \quad\ 3 \\ \hline \end{array}
\qquad
\begin{array}{r} 4\,4\,.1 \\ \times \quad\ 8 \\ \hline \end{array}
$$

2)
$$
\begin{array}{r} 4\,0\,.2 \\ \times\ 2\,.7 \\ \hline \end{array}
\qquad
\begin{array}{r} 5\,.2\,4 \\ \times\ 5\,.8 \\ \hline \end{array}
\qquad
\begin{array}{r} 6\,1\,7 \\ \times\ 8\,.6 \\ \hline \end{array}
\qquad
\begin{array}{r} 7\,0\,.4 \\ \times\ 0\,.9\,3 \\ \hline \end{array}
$$

3)
$$
\begin{array}{r} 2\,.1\,3 \\ \times 5\,6\,.5 \\ \hline \end{array}
\qquad
\begin{array}{r} 4\,7\,.5 \\ \times 2\,5\,.6 \\ \hline \end{array}
\qquad
\begin{array}{r} 9\,.7\,6 \\ \times 6\,8\,.8 \\ \hline \end{array}
\qquad
\begin{array}{r} 4\,.0\,1 \\ \times 1\,.0\,9 \\ \hline \end{array}
$$

4)
$$
\begin{array}{r} 2\,4\,0.2 \\ \times \quad\ 5.3 \\ \hline \end{array}
\qquad
\begin{array}{r} 7\,.0\,0\,4 \\ \times \quad\ 5\,.6\,7 \\ \hline \end{array}
\qquad
\begin{array}{r} 6\,7.1\,5 \\ \times\ 3\,.2\,3 \\ \hline \end{array}
\qquad
\begin{array}{r} 9\,0.0\,0 \\ \times\ 0\,.9\,5 \\ \hline \end{array}
$$

3.4) DIVISION OF DECIMALS:

A) The dividend has the decimal point.

	Step1: Put up the decimal point directly above it	Step 2: Divide Normally
$12\overline{)2.4}$	$12\overline{)24}$	$12\overline{)24}^{.2}$ $\underline{24}$

1) $2\overline{)5.8}$　　2) $3\overline{)18.6}$　　3) $4\overline{)0.180}$　　4) $2\overline{)1.73}$

5) $4\overline{)50.8}$　　6) $3\overline{)12.6}$　　7) $5\overline{)1.280}$　　8) $7\overline{)10.5}$

9) $2\overline{)0.158}$　　10) $9\overline{)18.9}$　　11) $8\overline{)1.80}$　　12) $6\overline{)1.74}$

13) $9\overline{)5.85}$　　14) $8\overline{)18.4}$　　15) $4\overline{)6.18}$　　16) $6\overline{)127.8}$

B) Division of Decimals: The divisor has the decimal point

Step1 • Get rid of the decimal in the divisor. • Add zeros to the dividend that are equal to how many numbers are on the right of the decimal.		Step 2 Divide Normally	**See?** There is only one number on the right of the decimal (1.2). So we add one zero only to the 24!
$1.2\overline{)24}$	$12\overline{)240}$	$\begin{array}{r} 20 \\ 12\overline{)240} \\ \underline{24} \end{array}$	

1) $0.2\overline{)58}$

2) $0.03\overline{)186}$

3) $4.2\overline{)84}$

4) $0.2\overline{)172}$

5) $0.04\overline{)2}$

6) $0.3\overline{)6}$

7) $0.06\overline{)3}$

8) $0.7\overline{)105}$

9) $0.2\overline{)15}$

10) $0.9\overline{)189}$

11) $0.8\overline{)36}$

12) $6.2\overline{)186}$

13) $0.1\overline{)5}$

14) $0.01\overline{)18}$

15) $0.01\overline{)6}$

16) $0.2\overline{)128}$

3.5) DECIMAL LANGUAGES: TERMINATING, NON-TERMINATING, REPEATING

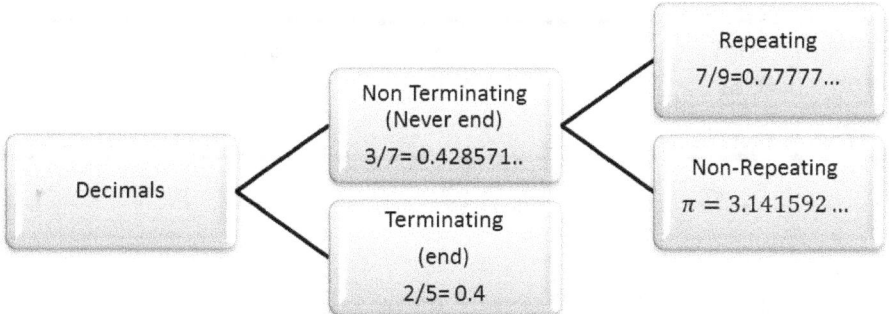

- **Terminating Decimals:** Decimals end up after several places.

  Examples: 1/5 = 0.2.   2/5 = 0.4   1/4 = 0.25

- **Non-Terminating Decimals:** Decimals that never end.

  Examples: **0.423423 423 .....**   **3.1459263.....**

- **Non-terminating decimals either repeat or do not repeat:**

  Repeating decimals have clear pattern of repeating digits: **0.423423...... 5.01616**

  The repeating part such as **423 or 16** is called **repetend.**

  To show that decimal is repeating put bar on the repetend. **0.$\overline{423}$ or 5.$\overline{016}$**

  Non-repeating decimals such as: $\pi = 3.145926358$ ....do not have a repetend and can go on forever.

Write as either repeating (R) or non-repeating (NR). Also, write a bar on the repetend:

1) 0.21212121

2) 7.425252525

7) 12.5175421..

3) 14.3333

4) 0.193193

8) 0.903050305

5) 4.0271717

6) 0.01010101

9) 12.74321054547

3.6) CHANGE TERMINATING DECIMALS INTO FRACTIONS

Get rid of the decimal. Multiply & divide with 10, 100, 1000..as needed and simplify	$0.5 \times \dfrac{10}{10} = \dfrac{5}{10} = \dfrac{1}{2}$	$1.14 \times \dfrac{100}{100} = \dfrac{114}{100} = \dfrac{57}{50}$

1) 0.3

2) 10.4

3) 1.02

4) 0.8

5) 0.65

6) 4.0

7) 7.05

8) 8.25

9) 9.5

10) 0.05

11) 0.174

12) 14.1

13) 4.33

14) 3.12

15) 11.5

3.7) CHANGE DECIMALS INTO PERCENT

To change Decimals into percent:	(a) 0.5 ↓	(b) 1.06 ↓
Multiply it by 100 and add the % sign.	$0.5 \times 100\% = 50\%$	106%

1) 1.3

2) 2.3

3) 1.02

4) 4.5

5) 6.5

6) 4.0

7) 7.05

8) 8.25

9) 9.5

10) 0.05

11) 0.174

12) 14.1

13) 4.33

14) 3.12

15) 11.5

UNIT 4: RATIOS, RATES AND PROPORTIONS

Perfect proportions!

The sphinx is a mystical structure with human head and lion's body. It is said that human head shows wisdom and the lion body shows strength. It was built by ancient Egyptians during the rule of the Pharaoh Khafra. Do you think it really existed? What do you think of the proportions of its parts?

4.1) RATIOS EXPLAINED

1. A ratio compares two quantities of the same kind.

Example 1, if there are 3 boys and 5 girls in a classroom we say:

"The ratio of boys to girls is 3 to 5. Or the ratio of girls to boys is 5 to 3".

2. Three ways to write ratios of boys to girls in the classroom:

✓   We can us use the word **to, colon, or a fraction:**

✓   The ratio of boys **to** girls is **3 to 5**   or 3 : 5   or $\frac{3}{5}$

3. If you write your answer as a fraction, write the first number at the top (numerator).

Example 2: (a).There are 7 boys and 3 girls in a classroom.

The ratio of girls to boys is $\frac{3}{7}$.

Example 3: There are 10 blue crayons and 6 red crayons.

The ratio of the red crayons to blue crayons is $\frac{6}{10} = \frac{3}{5}$

(Note we have simplified the fractions.).

4) **To see if two ratios are equal**: Check if their cross products are equal

Example 4: Write >, or < or = to compare these ratios:

a)	$\frac{1}{2} \bigcirc \frac{3}{6}$		b)	$\frac{6}{17} \bigcirc \frac{7}{19}$
	$\frac{1}{2} \bowtie \frac{3}{6}$			$\frac{6}{17} \bowtie \frac{7}{19}$
	6 × 1 = 2 × 3 or 6 = 6			6 × 19 7 × 17 *or* 114 < 119

4.2) RATIOS PROBLEMS: SIMPLIFY YOUR ANSWER

1. Jamal has a bag with 4 books, 8 pencils and 5 pens. What is the ratio of (a) books to pens? (b) Pens to pencils?

2. One in four students loves to play chess. In a class of 20 students, how many students love to play chess?

3. The foreign born persons in 2006 - 2010 were 7% in Minnesota and 13% for the US. What was the ratio of foreign born in Minnesota to that of the USA in 2006-2010?

(**Source:** US Census Bureau, Quick Facts, Minnesota, 2010)

Q 4 – 7: There are132 rooms in the White House including 35 bathrooms. There are 412 doors, 147 windows, 28 fireplaces, 7 staircases, and 3 elevators in the White House. (**Source:** www.whitehouse.gov)

4) What is the ratio of rooms to bathrooms?

5) What is the ratio of doors to windows?

6) What is the ratio of fireplaces to staircases?

7) What is the ratio of staircases and elevators to fireplaces?

8) Write equal (=) or greater (>) or less (<) for the following ratios:

(a) $\frac{3}{4}$(   )$\frac{3}{4}$ b) $\frac{15}{7}$(   )$\frac{20}{8}$ c) $2\frac{1}{5}$(   )$\frac{22}{10}$ d) 14(   )$\frac{28}{5}$

4.3) EQUIVALENT RATIO TABLES

Equivalent ratios are just like equivalent fractions:

$$\frac{2}{5} = \frac{6}{?}$$

Step 1) See that the 2 times 3 is what gives us the 6.	$\dfrac{2 ^{\times 3}}{5} = \dfrac{6}{?}$
Step2) So, multiply the 3 also to the 5 to get 15	$\dfrac{2^{\times 3}}{5_{\times 3}} = \dfrac{6}{15}$

Complete the ratio tables

1)

1	2
2	4
3	6
4	?

2	6
4	12
6	18
8	?

3	1
6	2
9	3
12	?

2)

0	0
5	20
6	24
7	?

5	6
10	12
15	18
?	24

3	2
9	6
27	18
?	48

3)

3	4
?	8
9	12
12	16

4	6
20	?
24	36
28	42

7	8
14	16
21	?
28	32

4)

8	9
16	?
24	27
32	36

1	3
2	?
?	9
4	12

8	9
?	36
56	45
64	?

4.4) UNIT RATES & UNIT COSTS

In most cases it is easy to understand prices, costs etc. when we know how much one unit such as 1 book, 1 gallon, 1 set of something costs. The rate of one unit is called **unit rate. When it is about cost** it is called **unit cost**.

Example1: 5 tea spoons of sugar gives 80 calories. Find the unit rate of **one teaspoon of sugar.**

$\dfrac{80 \text{ calories} \div 5}{5 \text{ teaspoon} \div 5} = \dfrac{16 \text{ cal}}{1 \text{ teaspon}}$	*Therefore, the unit rate is 16 calories per teaspoon of sugar.*

Example 2 dozens of eggs cost $6.50. What is the unit cost per egg?

$\dfrac{\$6.50}{24 \text{ eggs}} = \$0.27/egg$	*Therefore, the unit cost is 27 cents per gg.*

Find the Unit rates:

1) Typing 800 words in 12 minutes.

2) Find the heart rate if it is beating 420 betas in 6 minutes.

3) Travelling 240 miles in 6 hours

4) Scoring 60 points in 4 games.

Find the Unit cost:

5) $8 for 3 gallons of milk

6) 5 books for 75 dollars.

Find the total cost:

7) 10 gallons at $3.5 per gallon.

8) 48 pounds at 2.5 per pound.

9) **Which juice is cheaper?** A 320 milliliter bottle costing $ 4 or 600 milliliter bottle costing $8.4?

4.5) PROPORTIONS: EQUIVALENT RATIOS

Proportions is an equation that shows two ratios are equal: Let's find the missing number to complete the proportion:

$$\frac{5}{7} = \frac{?}{42}$$

Step 1) *See that the* **7** *times* **6** *is what gives us the* **42.* **Step2)** *So, multiply the 6 also to the 5 to get 30*	$\dfrac{5^{\times 6}}{7_{\times 6}} = \dfrac{30}{42}$

1) $\quad \dfrac{2}{4} = \dfrac{8}{?}$

2) $\quad \dfrac{2}{1} = \dfrac{?}{5}$

3) $\quad \dfrac{2}{3} = \dfrac{8}{?}$

4) $\quad \dfrac{2}{5} = \dfrac{?}{15}$

5) $\quad \dfrac{4}{10} = \dfrac{?}{20}$

6) $\quad \dfrac{4}{9} = \dfrac{?}{45}$

7) $\quad \dfrac{8}{20} = \dfrac{?}{5}$

8) $\quad \dfrac{21}{33} = \dfrac{?}{11}$

9) $\quad 15:18 = \underline{\quad}:6$

10) $\quad 1:3 = 8:\underline{\quad}$

11) $\quad 5:2 = 25:\underline{\quad}$

12) $\quad \dfrac{14}{49} = \dfrac{2}{?}$

13) $\quad \dfrac{7}{8} = \dfrac{28}{?}$

14) $\quad \dfrac{4}{7} = \dfrac{100}{?}$

15) $\quad \dfrac{14}{41} = \dfrac{56}{?}$

4.6) ARE TWO RATIOS EQUAL?

To find out if two ratios are equal use the **cross products**. See examples:

$\dfrac{5}{7} = \dfrac{15}{21}$	1) 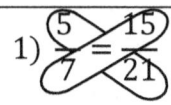	$5 \times 21 = 7 \times 15$ $105 = 105$

Decide if each pair of ratios are equal?

1) $\dfrac{2}{3}$, $\dfrac{8}{12}$

2) $\dfrac{2}{1}$, $\dfrac{6}{5}$

3) $\dfrac{3}{8}$, $\dfrac{1}{4}$

4) $\dfrac{56}{90}$, $\dfrac{7}{15}$

5) $\dfrac{10}{15}$, $\dfrac{3}{20}$

6) $\dfrac{4}{9}$, $\dfrac{?}{45}$

7) $\dfrac{8}{20}$, $\dfrac{48}{120}$

8) $\dfrac{28}{44}$, $\dfrac{7}{11}$

9) $\dfrac{16}{27}$, $\dfrac{3}{5}$

10) $\dfrac{36}{40}$, $\dfrac{9}{10}$

11) $\dfrac{9}{8}$, $\dfrac{7}{6}$

12) $\dfrac{12}{51}$, $\dfrac{4}{17}$

14) To make sambusa you will need 3 cups of flour for every 21 sambusa. How many Sambusa could be prepared from 9 cups of flour?

13) The ratio of winning to losses of a basketball team is 5 to 7. Which could be the teams winning record?

a) 10 wins and 16 losses

b) 15 wins 20 losses

c) 10 wins and 15 losses

d) 25 wins and 35 losses

4.7) SOLVE PROPORTIONS

To solve proportions easily use the cross products. This is how: See Examples

1) $\dfrac{5}{6} = \dfrac{y?}{24}$ $y = \dfrac{5 \times 24}{6} = 20$

2) $\dfrac{n?}{8} = \dfrac{9}{24}$ $n = \dfrac{8 \times 9}{24} = 3$

✓ See: we show the missing value with a letter called variable such as x, y, n, etc.
✓ Also, the number that is opposite (facing) the variable becomes the denominator. In the examples the 6 and the 24 were across (opposite) the variables y and n.

Practice: Solve for the variable:

1) $\dfrac{3}{4} = \dfrac{y}{20}$ 2) $\dfrac{2}{10} = \dfrac{x}{5}$ 3) $\dfrac{n}{16} = \dfrac{9}{4}$

4) $\dfrac{2}{5} = \dfrac{x}{15}$ 5) $\dfrac{4}{y} = \dfrac{36}{81}$ 6) $\dfrac{4}{k} = \dfrac{24}{42}$

7) $\dfrac{8}{w} = \dfrac{72}{18}$ 8) $\dfrac{10}{11} = \dfrac{x}{22}$

9) If a car travels 180 miles for 3 hours, how long it takes it to travel for 720 miles. Show the proportion and solve.

10) During a field trip, 8 teachers are needed for every 40 students. How many teachers are needed for 600 students? Show the proportion and solve.

11) Asha can type 120 words in 3 minutes. Which proportion can be used to find the number of minutes she can type 400 words. Use **m** for minutes

a) $\dfrac{m}{120} = \dfrac{400}{3}$ b) $\dfrac{120}{3} = \dfrac{400}{m}$ c) $\dfrac{m}{120} = \dfrac{3}{400}$ d) $\dfrac{120}{3} = \dfrac{m}{400}$

UNIT 5: PERCENT, FRACTIONS AND DECIMALS RELATIONSHIPS

5.1) What is Percent?

5.2) Percent Concept Graphics

5.3) Change Fractions into Percent

5.4) Change Mixed Numbers into Percent:

5.4) Change Percent into Fractions

5.5) Change Mixed Number Percent into Fractions

5.6) Change Percent into Decimals

5.7) Change Decimal Percent into Fractions

5.8) Mixed Review: Fractions, Decimals, Percent

VERY COLORFUL PERCENT!

Did you know that the tail feathers of peacocks, spread more than 60 percent of the bird's total body? Why the bird needs 60% of the total body of the bird to be covered by beautiful and colorful feathers? Remember that the male is called peacock and the female is called peahen. Together they are called peafowl.

5.1) WHAT IS PERCENT?

> The word percent means "out of a hundred" or a part of a hundred. For example if you split with your friend $100, each takes $50 out of the $100 or **50 percent**. Instead of writing every time "Percent" we simply use the percent sign %.
>
> **So, 50 percent is written as 50 %.**

Common Sense Percent Problems:

Write the following as a percent. Use common sense and your knowledge of fractions and the examples above:

1) You split with three other friends $100. What percent will each get?

2) In every four free throws, Shadiya makes three. What percent Shadiya scores?

3) Ten brothers and sisters shared $800. What percent will each get?

4) Awil was the fourth in the math tests in a class of 10 at Pre-University. What percent of students were lower? What percent were higher?

From problems 5 to 7 use the pie chart:

5) Which two activities does Dalmar spends 50% of his time?

6) Estimate the percent of time Dalmar spends on playing.

7) What are the total percent of time Dalmar spends on all four subjects: Writing, reading, math and science?

8) Ouch! 75% of all accidents happen within 5 miles of home. Could you explain what does that mean?

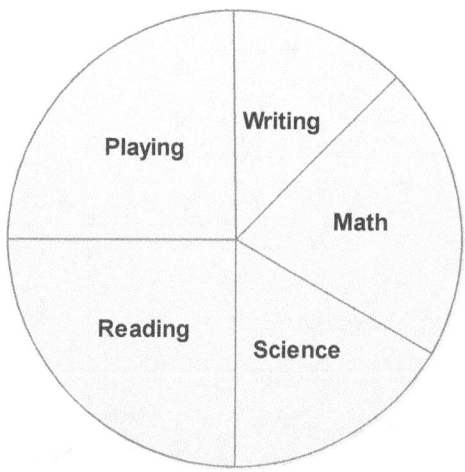

Dalmar's Activity Chart

5.2) PERCENT CONCEPT GRAPHICS:

What percentage of the shape (a) is shaded? B) Not shaded

1) Shaded: 34% Not Shaded: 68%	2) Shaded:_____ Not Shaded:_____	3) Shaded:_____ Not Shaded:_____
4) Shaded:_____ Not Shaded:_____	5) Shaded:_____ Not Shaded:_____	6) Shaded:_____ Not Shaded:_____
7) Shaded:_____ Not Shaded:_____	8) Shaded:_____ Not Shaded:_____	9) Shaded:_____ Not Shaded:_____
10) Shaded:_____ Not Shaded:_____	11) Shaded:_____ Not Shaded:_____	12) Shaded:_____ Not Shaded:_____

5.3) CHANGE FRACTIONS INTO PERCENT

To change fractions into percent: Multiply it by 100 and add the % sign.	(a) $\frac{2}{5}$ $\frac{2}{5} \times 100\%$ $= \frac{2 \times 100}{5} = 40\%$	(b) $\frac{1}{4}$ $\frac{1}{4} \times 100$ $\frac{1 \times 100}{4} = 25\%$

Change the following fractions into percent:

1) $\frac{3}{4} =$ 2) $\frac{4}{5} =$ 3) $\frac{1}{4} =$ 4) $\frac{3}{5} =$

5) $\frac{1}{5} =$ 6) $\frac{4}{5} =$ 7) $\frac{5}{6} =$ 8) $\frac{1}{20} =$

9) $\frac{1}{12} =$ 10) $\frac{1}{14} =$ 11) $\frac{5}{12} =$ 12) $\frac{1}{8} =$

13) $\frac{1}{24} =$ 14) $\frac{3}{50} =$ 15) $\frac{1}{60} =$ 16) $\frac{2}{25} =$

Why teens check their online more than 100 times a day? According to CNN research (October 13, 2015);

✓ 61% of teens said they wanted to see if their online posts are getting likes and comments.
✓ 36% of teens said they wanted to see if their friends are doing things without them.
✓ 21% of teens said they wanted to make sure no one was saying mean things about them

5.4) CHANGE MIXED NUMBERS INTO PERCENT:

A) Method 1

1) Change the mixed numbers into fractions	$2\frac{1}{2} = \frac{5}{2}$
2) Multiply it by 100%	$\frac{5}{2} \times 100\% = \frac{500}{2}$
3) Simplify the fractions	$\frac{500}{2} = 250\%$

1) $3\frac{1}{2} =$

2) $5\frac{1}{2} =$

3) $7\frac{1}{4} =$

4) $8\frac{3}{4} =$

5) $9\frac{1}{5} =$

6) $5\frac{2}{5} =$

7) $7\frac{4}{5} =$

8) $14\frac{1}{2} =$

9) $11\frac{1}{4} =$

10) $2\frac{1}{20} =$

11) $12\frac{3}{10} =$

12) $17\frac{1}{100} =$

B) Method 2: Change mixed numbers into percent

1) Change the mixed numbers into decimals	$2\frac{1}{2} = 2 + 0.5 = 2.5$
2) Multiply it by 100%	$2.5 \times 100 = 250\%$

1) $3\frac{1}{2} =$

2) $5\frac{1}{2} =$

3) $7\frac{1}{4} =$

4) $8\frac{1}{2} =$

5) $8\frac{3}{4} =$

6) $9\frac{1}{5} =$

7) $5\frac{2}{5} =$

8) $12\frac{3}{5} =$

9) $7\frac{4}{5} =$

10) $14\frac{1}{2} =$

11) $11\frac{1}{4} =$

12) $3\frac{8}{10} =$

13) $2\frac{1}{20} =$

14) $12\frac{3}{10} =$

15) $17\frac{1}{100} =$

16) $9\frac{1}{80} =$

5.4) CHANGE PERCENT INTO FRACTIONS

Example: Change the percent into fraction: (a) 3%,   b) 12%

1) Just divide it by 100 & get rid of the percent sign	$\dfrac{3}{100}$	$\dfrac{12}{100}$
2) Simplify as needed	$\dfrac{3}{100}$	$\dfrac{12 \div 4}{100 \div 4} = \dfrac{3}{25}$

1. 13% =

2) 40% =

3) 17% =

4) 45% =

5) 200% =

6) 41% =

7) 24% =

8) 16% =

9) 55 % =

10) 11% =

11) 42% =

12) 28% =

13) 4% =

14) 54% =

15) 28% =

5.5) CHANGE MIXED NUMBER PERCENT INTO FRACTIONS

Example: Change the percent into fraction: $2\frac{1}{2}\%$

1) **Change** any mixed numbers into fractions	$\dfrac{5}{2}$
2) **Multiply** the lower number by 100	$= \dfrac{5}{2 \times 100} = \dfrac{5}{200}$
3) **Simplify** the fractions	$= \dfrac{5 \div 5}{200 \div 5} = \dfrac{1}{40}$

1. $2\frac{1}{4}\% =$　　　　　2. $4\frac{1}{2}\% =$　　　　　3. $5\frac{1}{4}\% =$

4. $12\frac{1}{5}\% =$　　　　5. $1\frac{3}{5}\% =$　　　　6. $6\frac{2}{5}\% =$

7. $8\frac{1}{4}\% =$　　　　8. $8\frac{1}{2}\% =$　　　　9. $12\frac{2}{5}\% =$

10. $8\frac{3}{4}\% =$　　　　11. $3\frac{1}{5}\% =$　　　　12) $15\frac{2}{5}\% =$

Who enjoys social media more?

44% of teen girls say they enjoy it "a lot" versus 29% for boys (Source, CNN, October, 13,2015)

5.6) CHANGE PERCENT INTO DECIMALS:

A) Method 1

Examples: Change the percent into Decimals: (a) $12\frac{1}{2}\%$ (b) 3%

1. Change any mixed numbers into improper fraction.	$\frac{25}{2}\%$	3%
2. Multiply it by 0.01 and remove the % sign.	$\frac{25}{2} \times 0.01$ $= 0.125$	3×0.01 $= 0.03$

1) 13%   2) 40%   3) 17%

4) 2.5%   5) 1.3%   6) 45%

7) 200%   8) 0.2%   9) $4\frac{1}{2}\%$

10) 0.75%   11) 40 %   12) 0.02%

 *Did you know that the human brain is 80% water?*

B) Method 2: change Percent into Decimals

To change percent into decimals :	(a) 25%	(b) 206
Divide the percent by 100 and get rid of the % sign.	$\dfrac{25}{100} = 0.25$	$\dfrac{206}{100} = 2.06$

1) 13%

2) 2.3%

3) 1.02%

4) 5.3%

5) 45%

6) 16.5%

7) 4.07%

8) 6.28%

9) 7.05%

10) 8.25%

11) 9.5%

12) 4.5%

13) 0.15%

14) 574%

15) 14.1%

16) 15%

17) 4.33%

18) 12%

19) 125%

20) 0.5%

Wow! About 70 percent of an adult's body is made up of water?

5.7) CHANGE DECIMAL PERCENT INTO FRACTIONS

Example1: Change the percent into fraction: (a) 0.5%   b) 0.24%

1) Divide it by 100. Rid of % sigh	$\dfrac{0.5}{100}$	$\dfrac{0.24}{100}$
2) Get rid of the decimal. Multiply & divide with 10, 100,1000...as needed	$\dfrac{0.5}{100} \times \dfrac{10}{10} = \dfrac{5}{1000}$	$\dfrac{0.24}{100} \times \dfrac{100}{100} = \dfrac{24}{10000}$
3) Simplify the fractions	$= \dfrac{5}{1000} = \dfrac{1}{200}$	$\dfrac{24}{1000} = \dfrac{3}{125}$

1) 1.3% =

2) 0.4% =

3) 1.7% =

4) 4.5% =

5) 2.0% =

6) 4.1% =

7) 0.22% =

8) 1.6% =

9) 5.5 % =

10) 1.1% =

11) 4.5% =

12) 2. 8%

Discussion: Scary Percent: *According to a study reported in Washington Post in 2010, 28% of traffic accidents occur when people talk on cellphones or send text messages while driving!*

5.8) MIXED REVIEW: FRACTIONS, DECIMALS, PERCENT

Fill in the blanks with the correct fraction, percent or decimal: Show your work!

	Fraction	Decimal	Percent
1)	$\dfrac{2}{5}$	0.4	40%
2)	$\dfrac{1}{5}$		
3)		0.7	
4)			12%
5)	$\dfrac{3}{4}$		
6)		21.4	
7)			1.5%
8)	$\dfrac{5}{8}$		
9)		5.4	
10)	$12\dfrac{1}{5}$		
11)			13.8%
12)	$12\dfrac{1}{5}$		

UNIT 6: PERCENT PROBLEMS

Do you thing the lion share is fair?

The Lion went once hunting along with the Fox, the Jackal, and the Hyena. Then came the question how the hunt should be divided. The fox said I know: The lion takes 25% for being the king of beasts. Another 25% also for being the wise. The third quarter is also for the lion for taking part in the chase. Let's just talk about how to share the last quarter. The lion laughed and said "Fox, I admit. You're the best judge". Do you agree or disagree with the lion?

6.1) PERCENT: FINDING THE PART

Hint: Remember by **base** we mean the whole or the big number.

Example: What is 6% of 300?

$Part = Percent\ (in\ decimal) \times Base$ $\qquad Part = 0.06 \times 300 = 50$

✓ Remember of means *times* ($\times$)

1) What is 2% of 200?

2) What is 4% of 300?

3) What is 15% of 500?

4) What is 0.25% of 62?

5) What is 8.2 % of 500?

6) What is 7% of 6.54?

7) What is 21% of 800?

8) What is 22% of 1600?

9) What is 0.05% of 1000?

10) What is 12% of 700?

6.2) FINDING THE PERCENT

Hint: Remember the part is always less than the base (or the whole):

What percent is $10 out of $120?   (Or $10 is what percent of $120?)
$Percent = \dfrac{Part}{Base} \times 100$ $Percent = \dfrac{10}{120} \times 100 = 50\%$

1) What percent is $2 out of $50?

2) $4 is what percent of $56?

3) 2 is what percent of 8?

4) What percent is 4 out of 60?

5) What percent is 24 out of 504?

6) What percent of 400 is 80?

7) What percent is 20 out of 800?

8) $70 is what percent $560?

9) What percent of 350 is 7?

10) What percent of 12 is 4?

6.3) FINDING THE BASE

25 % of what number is 40? *You can do it in two ways:*	
$Base = \dfrac{part}{Percent\ (in\ decimals)}$ $= \dfrac{40}{0.25} = 160$	$Base = \dfrac{part}{Percent} \times 100$ $= \dfrac{40}{25} \times 100 = 160$

1) 18% of what number is 54?

2) 50% of what number is 1.5?

3) 60 is 20% of what number?

4) 2.8 is 50% of what number?

5) 2% of what number is 10?

6) 95 is 45% of what number?

7) 5% of what number is 50?

8) 280 is 14% of what number?

9) 10% of what number is 47?

10) 210% of what number is 525?

6.4) FIND THE PERCENT CHANGE

Example: The price of a shirt decreased from $20 to $14. What is the percent of change?

$$\textbf{Percent Change} = \frac{|Original\ amount - new\ amount|}{Original} \times \textbf{100\%}$$

$$\text{Percent Change} = \frac{|20 - 14|}{20} \times 100\% = \frac{6}{20} \times 100\% = 30\%$$

Important: the little bars | | in the formula is the absolute value. That means the answer should always be positive!

1) The price of math books increased from $10 to $12 this week end. What is the percent change?

2) On the New Year's Eve the price of shoes decreased from $30 to $20. What is the percent change?

3) Kadar's monthly income for the last year was $2500. This year it is $2900. What is his percentage increase?

4) The speed of car increased from 40 mph to 55 mph. What is the percent increase?

5) Last year, the total number of students of the new school was 200. This year it has increased to 350. What is the percent increase?

6) Siman was 66 inches tall before her height increased by 4 inches. What is the percent change in her height?

7) Farah's weight on January was 152 pounds. This month he lost 12 pounds. What is the percent change?

8) Ahmed practices math for 12 hours a week. Now he is planning to increase it to 16 hours. What will be the percent change?

6.5) PERCENT: MIXED PRACTICE

1) What is 5% of 200?

2) 4% of what number is 30?

3) $20 is what percent of 500?

4) What is 25% of 800?

5) 75 is what percent of 600?

6) What is 13% of 90?

7) What percent of 620 is 248?

8) What is 12% of 1600?

9) 15% of what number is 12?

10) What is 0.052% of 8100?

Too Much Time online: 92% of teens report going online daily including 40% who said are online all the time (Pew Research Center, April 9, 2015)

UNIT 7: CONSUMER MATH: TAXES, DISCOUNTS AND INTEREST RATES

7.1) Calculate the Tax/ Other increases

7.2) Calculate the Discounts/ Other Decreases

7.3) The Simple Interest

7.4) Compound Interest

The highest Growth Rate?

Sauropod dinosaurs were the largest terrestrial animals. It has been estimated that *Apatosaurus* reaches its adult body mass of about 25,000 kg in just 15 years, with a maximum growth rate over 5000 kg/yr. could you estimate your yearly growth?

7.1) CALCULATE THE TAX/ OTHER INCREASES

To calculate the tax part or other increases: Multiply the tax percent with the amount:

Example 1: How much tax you will pay for a shirt costing $40 with 10% tax? $$0.10 \ (\$40) = 4$$	What is the total cost of shoes costing $45 before a 8% tax: $$\$45 + 0.08 \ (\$45) = \$45 + 3.6 = 48.6$$

1) What is $400 increased by 20%?

2) The cost of a car was $15,000 before adding 8% tax. What is the total cost?

3) Amal's monthly income for last year was $2500 before it increased by 10%. What is her new monthly income?

4) The number of students for the new school was 300. This year it has increased by 12%. How many students are there now?

5) Farah was 60 inches tall before his height increased by 4%. What is his new height?

6) Awale practices biology for 12 hours a week. Now he is planning to increase it by 18%. What will be the new number of hours for practice?

7.2) CALCULATE THE DISCOUNTS/ OTHER DECREASES

Example: The price of a shirt was $20 but is now decreased by 15%. What is the price now?

$20 - 0.15 ($20) = $17

1) In December, the price of shoes decreased by 25%. The original price was $36. What is the new price?

2) Leyla's weight on January was 152 pounds. This month she lost 12% of her weight. What is her weight now?

3) The population of small town was 570. This year 11% of the residents have moved. How many people now live in the small town?

4) The original price of the computer game was $30 before it was discounted by 15%, what is the sale price now?

5) The new scooter cost $500. There are two offers. Either get a discount of $50 or 11% discount. Which option is better for a buyer?

6) Tom was not working well on his math lately. His grade decreased by 24%. His original grade was 95. What is his grade now?

7.3) THE SIMPLE INTEREST

Definition:

Interest is the amount of money paid for the use of money some body borrows (called the **principal)** from a bank or other financial institutions. **Look at the example to understand how to calculate it.**

Example: Warsame has $120 in saving account that earns 8% interest annually. How much will he have in 2 years?

$$Interest = Principal \times rate \times Time$$

$$I = PRT = \$120 \times 0.08 \times 2 = \$19.2$$

In 2 years Warsame will have = Original amount + Interest = 120+ 19.2= 139.2

*****Remember **T** is time in years and the **P** (principal) is the starting amount

Find the interest earned on each account:

1) $870 at 6% simple interest for 4 years.

2) $2500 at 7% interest for 10 years.

Find the balance in each account:

3) $ 756 invested at 5% simple interest for 12 years.

4) Omar borrowed $800 for 5 years loan from a bank to pay for his studies. The bank charges a simple interest of 3%. How much does Omar need to repay after 5 years?

5) Asha has $700 in an account that earns 5%. How much money she will have in eight years?

6) A business women deposits $8000 in a bank that pays 9% simple interest. What will be the balance after 5 years?

7) **Challenge:** Halima deposited $2000 in a bank that pays 5% simple interest. How many years will it take for the balance to be $2500?

7.4) COMPOUND INTEREST

Definition:

A compound Interest is the interest that is paid in the principal and any interest that is added or left in the account. **Look at the example to understand how to calculate it.**

Example: Nasra deposited $150 in saving account that pays 5% compound interest. How much money Nasra has after 3 years?

$$Balance = Principal\ (1 + rate)^{Time}$$

$$Balance = \$150(1 + 0.05)^3 = 173.64$$

After 3 years Warsame will have = $173.64

Find the balance on each compound interest account:

1) $870 at 6% for 4 years.

2) $7500 at 8% interest for 10 years.

3) $ 8056 invested at 5% for 8 years.

4) Tamara has $900 in an account that earns 10% compound interest. How much money she will have in eight years?

5) A business man deposits $6000 in a bank that pays 9% compound interest. What will be the balance after 5 years?

6) **Challenge:** Ali deposited $2000 in a bank that pays compound interest for 2 years. The balance after 2 years was $2400? What was the interest rate?

UNIT 8: MEASUREMENTS

How long could you be sitting to study?

Camels are known for their **endurance**. They can live in the dessert for many days without food or water. But, do you know a thirsty camel can drink up to 135 liters in a single sitting. Can you guess how many gallons are those? Figure it at the end of the unit!

8.1) MEASUREMENTS 1: THE BASIC UNITS

Who wears the largest shoes in the NBA players? Well, It is said that LeBron James wears size 14 (inches), Yao Ming wears size 18 inches while, Shaquille O'Neal', wears size 22 (inches), (source: Solecollector.com)

So, if there were no units of measure for shoes, milk, clothes, buyers and sellers would be very confused. Sure, Shaq would have gone from store to store and try the right size. That would be a lot of work!

The U.S measures and the Metric System

There are two systems of measurements: The **Metric Measures** used in most of the world and **U.S Measures** used mainly in the U.S and Canada. Listed below are most common units of length, weight, and volume. **Also don't forget the abbreviations.** Math people use these abbreviations often.

The U.S measure		The Metric Measures
inches (in) feet(ft) yards (yd) miles (mi)	Length	millimeter (mm) centimeter (cm) meter (m) kilometer (km)
ounces (oz) pounds (lb), tons (t),	Weight	milligram(mg) centigram (cg) gram (g) kilogram (kg)
fluidounces (fl oz) cups (c) pints (pt) quarts (qt), gallons (gal)	Volume	milliliters (ml) liters (l) cubic centimeter (cm^3) cubic meter (m^3)

8.2) US MEASURES: UNITS OF LENGTH

12 in = 1 ft	3ft = 1 yd	1760 yd = 1 mi	5280 ft = 1 mi

Example problems	Relationships from table above	Multiply with the right conversions so that the units cancel out
1) Convert 20 feet. into inches	1ft = 12 in	$20\,ft \times \dfrac{12\,in}{1\,ft} = 240\,in$
2) Covert 5280 yards into miles.	1760 yd = 1mi	$5280\,yd \times \dfrac{1\,mi}{1760\,yd} = \dfrac{5280\,mi}{1760} = 3\,mi$
2) Covert 144 inches into yards	12 in = 1ft 3in = 1yd	$144\,in \times \dfrac{1\,ft}{12\,in} \times \dfrac{1\,yd}{3\,ft} = 4\,yds$

The Problem: Convert	The Relationships You need	Multiply with the right conversions so that the same units cancel out
1) 15 ft. into in.	12 in = 1 ft.	
2) 144 inches into feet.	$12\,in = 1ft$	
3) 72 ft. into yd.		
4) 9 ft. into yd.		
5) 3.5 mi into ft.	1 m = 39.4 in	
6) 2640 ft. into mi.		
7) 120 in into yd.		
8) 3 mi into feet		
9) 8800 ft. into miles		
10) Safiya walked for 2 miles to the school. Express that into feet		

8.3) METRIC SYSTEM: UNITS OF LENGTH

1 km = 1000 m	1 m = 100 cm	1 m = 1000 mm	1 cm = 10 mm

Example problems	Relationships from table above	Multiply with the right conversions so that the units cancel out
1) Convert 5 km. into meters	1km = 1000 m	$5 \, \cancel{km} \times \dfrac{1000 \, m}{1 \, \cancel{km}} = 5000 \, m$
2) Covert 5200 mm into meters.	1000 mm = 1m	$5200 \, \cancel{mm} \times \dfrac{1 \, m}{1000 \, \cancel{mm}} = 5.2 \, m$
2) Covert 24000 cm into km.	100 cm = 1m 1000m = 1km	$24000 \, \cancel{cm} \times \dfrac{1 \, \cancel{m}}{100 \, \cancel{cm}} \times \dfrac{1 \, km}{1000 \, \cancel{m}} = 0.24 \, km$

The Problem: Convert	The Relationships You need	Multiply with the right conversions so that the same units cancel out
1) 15 m. into cm.	1m = 100 cm.	
2) 1200 mm into cm.		
3) 2.2 m into mm.		
4) 1.2 km into m.		
5) 36000 m into km.	1 km = 1000 m	
6) 1.2 km into m.		
7) 0.25 km into mm.		
8) 360,000 mm into km		
9) 8800 m into km		
10) Vicky walked for 2 km to the school. Express that into meters.		

8.4) CONVERSION BETWEEN METRIC & U.S MEASURES OF LENGTH

1 in = 2.5 cm	1 m = 39.4 in	1 mi = 1.61 km	1 km = 0.62 mi

Example problems	Relationships from table above	Multiply with the right conversions so that the units cancel out
3) Convert 5 km. into miles	1km = 0.62 mi	$5\ km \times \dfrac{0.62\ mi}{1\ km} = 3.1\ mi$
2) Covert 500 cm into inches.	2.5 cm = 1in	$500\ cm \times \dfrac{1\ in}{2.5\ cm} = 200\ in$
4) Covert 64000 in into km.	1m = 39.4 in 1000m = 1km	$64000\ in \times \dfrac{1m}{39.4\ in} \times \dfrac{1\ km}{1000\ m} = = 1.62\ km$

The Problem	The Relationships You need	Multiply with the right conversions so that the same units cancel out
1) 5 m into inches.	1m = 39.4 in.	
2) 150 inches into meters.	1m = 39.4 in.	
3) 6.2 inches into cm.		
4) 25 cm into inches.	2.5 cm = 1in	
5) 640 miles into km.	1 km = 0.62 mi	
6) 7.2 km into miles.		
7) 0.25 m into inches.		
8) 860 inches into meters	1 in = 2.5 cm 1m =100 cm	
9) 25 m into inches		
10) Guled walked for 2 miles to home. Express that into meters.	1 mi = 1.61 km 1 km = 1000 m	

8.5) UNITS OF WEIGHT

Now that you are familiar with how to build and use the conversion systems, you have to be able to apply it quickly in converting between units of weight.

A) The U.S Measures	B) The Metric measures	C) Conversion between U.S and Metric measures
1lb = 1 6 oz 1T = 2000 lb	1 kg = 1000 g 1 g = 1000 mg	1 lb = 454 g 1 kg = 2.2 lb 1 ton = 907.2 kg

The Problem: Convert	The Relationships You need	Multiply with the right conversions so that the units cancel out!
1) 12 pounds into ounces	1 lb = 16 oz	$12lb \times \dfrac{16\ oz}{1lb} = 192\ oz$
2) 0.2 Kg to ounces.	$1kg = 2.2\ lb$ $1\ lb = 16\ oz$	$0.2\ kg \times \dfrac{2.2\ lb}{1\ kg} \times \dfrac{16\ oz}{1lb} = 7.04\ oz$
3) 4000 g to kg.	1 kg = 1000 g	
4) 22 lb into g.		
5) 0.25 T into g		
6) 15 g into milligrams		
7) 20 lb into g		
8) 1 T into kg.		
9) 2400 grams into pounds.		
10) 360 oz into pounds		

8.6) UNITS OF CAPACITY OR VOLUME

My friend visited us from Denmark one time. Anytime I buy few gallons of fuel, he can't help but to compare prices in Europe with that of U.S. How many liters are five gallons? After I give my· estimate he says, it is cheaper in the U.S. These relationships below helped me for the estimates:

The U.S Measures		The Metric measures	Conversion between U.S and Metric measures
8 fl oz = 1 cup 2 c = 1 pt 2pt = 1 qt	1 c = 0.25 qt 1 gal = 4 qt 1 qt = 32 fl oz	$1 L = 1000\ ml$ $1ml = 1\ cm^3$ $1000\ l = 1\ m^3$	$1 L = 1.06$ qt $1qt = 946$ ml $1 c = 0.24\ l$

The Problem: Convert	The Relationships You need	Multiply with the right conversions so that the units cancel out
1) 8 cups into quart	1 c = 1 pt 2 pt = 1 qt	$8\ c\ \times\ \dfrac{1\ pt}{1\ c}\ \times\ \dfrac{1\ qt}{2\ pt} = 4$ qt
2) 2 L into gallons.	1L = 1.06 qt. 4 qt = 1 gallon =	$2\ L\ \times\ \dfrac{1.06\ qt}{1\ L}\ \times\ \dfrac{1\ gal}{4\ qt} = 0.8$ gal
3) 4000 ml into L.		
4) 2 gal into pt.		
5) 25 c into L		
6) 10 L into fl oz		
7) 4 pints into milliliters		
8) 20 gal into ml		
9) 473 ml into fl oz	1qt = 946 ml 1 qt = 32 fl oz	
10) 1000 cups into ml.		

8.7) ADD AND SUBTRACT MIXED UNITS

There is an African proverb that says "oil and water cannot be mixed". When adding and subtracting units of measures such as feet and inches, pounds and ounces, make sure that you change them to the same units. Never try to mix oil and water. It won't happen!

<u>Example 1</u>: Add: (a) 12 lb 15 oz. + 2lb 6 oz.   (b) 6ft 8 in + 3 ft 10 in

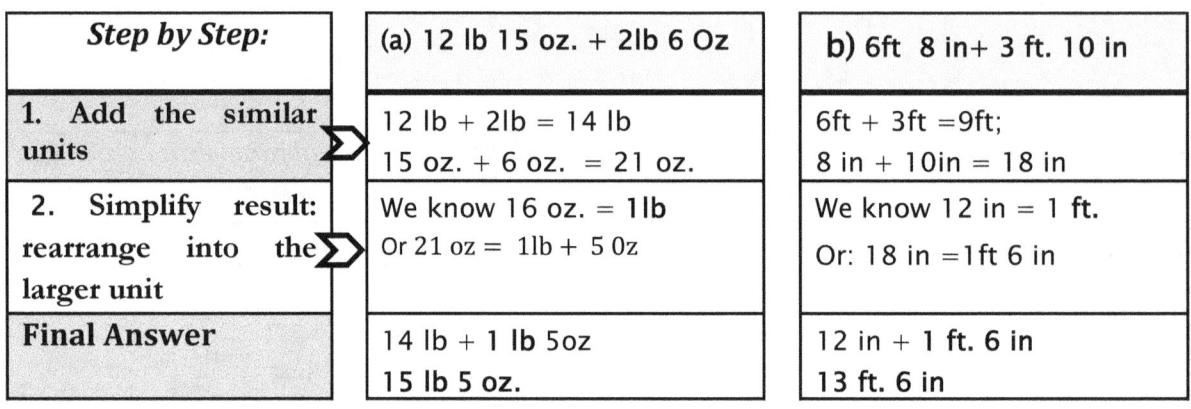

Step by Step:	(a) 12 lb 15 oz. + 2lb 6 Oz	b) 6ft 8 in+ 3 ft. 10 in
1. Add the similar units	12 lb + 2lb = 14 lb 15 oz. + 6 oz. = 21 oz.	6ft + 3ft =9ft; 8 in + 10in = 18 in
2. Simplify result: rearrange into the larger unit	We know 16 oz. = 1lb Or 21 oz = 1lb + 5 0z	We know 12 in = 1 ft. Or: 18 in =1ft 6 in
Final Answer	14 lb + 1 lb 5oz **15 lb 5 oz.**	12 in + 1 ft. 6 in **13 ft. 6 in**

<u>Example 2</u>: Subtract (a) $12\,ft\,5in - 7ft\,3in$ (b) $12\,ft\,5in - 7ft\,9in$

Subtractions could be either straight forward (Example 2a) or require borrowing (Example 2b)

Step by Step:	(a) 12 ft 5 in - 7ft 3 in	(b) $12\,ft\,5in - 7ft\,9in$
✓ **Subtract similar units.** ✓ **Borrow if you need to**	$12ft - 7\,ft = 5ft$ $5\,in - 3in\ = 2\,in$ Final answer: <u>5ft 2in</u>	We can't subtract 9 from 5. So, borrow 1ft from the 12 ft and change it to inches 11 ft 17 in − 7ft 9 in = <u>4 ft 8 in</u>

Practice of Adding or Subtracting Units:

	a)
1)	9 ft 4 in. + 7 ft 9 in
2)	9 m 20 cm − 7 m 30 cm
3)	13g 200 mg − 4 g 900 mg
4)	15 L 500 ml + 7L 600 ml
5)	8 ft 2 in. − 2 ft 10 in

	b)
1)	10 *yd* 5*ft* −7yd 8ft
2)	1 gal 2 qt. +7 gal 4qt.
3)	8c 2fl. oz. −2c 5 fl oz.
4)	9 km 400 m − 7 km 800 m
5)	8 *yd* 1*ft* −4yd 2 ft

	c)
1)	7 *m* 70 *cm* + 5m 40 cm
2)	12 kg. 200g − 8 kg. 700g
3)	4 ft 1 in. − 2 ft 9 in
4)	3 *lb* 2 oz. −2 *lb* m 3 oz
5)	10 *m* 10 *cm* − 5m 40 cm

UNIT 9: GEOMETRY

What is Geometry?

Geometry is the study of the size, shape and position of figures. The word geometry itself is a Greek word and it means "earth measure"

Plane Geometry includes lines, triangles, circles and polygons drawn on a flat surface called plane.

Solid geometry includes three dimensional figures (3-D) such as cubes, cylinders cones, spheres, prisms, and pyramids.

Other types of geometry include analytic geometry, spherical geometry etc.
In this Unit we will focus mainly on plane Geometry!

Did you know that the four sides of the Taj mahal are perfectly identical!

Parallel

The four sides of the Taj Mahal are perfectly identical

9.1) GEOMETRY: SOME BASIC DEFINITIONS

A Point is a dot in space that has no size and no areas. It is named with a capital letter such as a "D".

A Line: Two or more points that are connected form a line that extends to both directions indefinitely. A line has no thickness. It is read either as $\overleftrightarrow{AB}$ or as $\overleftrightarrow{BA}$ **or just line AB or line BA**

A Line Segment has two endpoints. It is read as $\overline{AB}$ or $\overline{BA}$ or just segment AB etc.

A Ray goes only one direction and has one end point. It is read as $\overrightarrow{AB}$ or just ray AB.

Parallel lines: Two lines that never cross each other.

Perpendicular lines: Two lines that cross each other at 90°

Polygon: A closed figure made of line segments each of which intersects with exactly two other line segments.

Quadrilateral is a four sided polygon

Parallelogram is a four **sided** polygon with two pairs of parallel sides.

9.2) HOW TO NAME ANGLES

Angle: Two line segments or rays that share a common endpoint form an angle.

Vertex: Where the line segments meet is called Vertex. "B" is the vertex.

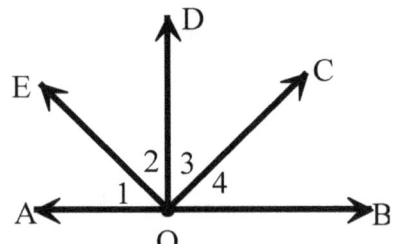

Angles are named by reading their labels: we can name the angles in three different ways:

∠ABC (read as Angle ABC) or ∠CBA or simply ∠B. You see the "B" is always in the middle in all cases.

Prcatice: Use the figure on the right: Name each angle in two ways

1) <1 _____

2) <2 _____

3) <3 _____

4) <4 _____

Name each angle indicated in three ways.

5)

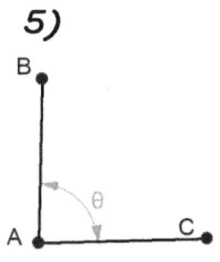

6)

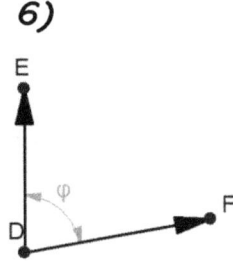

7)

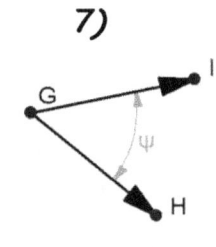

9.3) HOW TO MEASURE ANGLES:

A **Protractor** is used to read the angle measurements.

> Angles are measured usually through degrees (°).
> A full circle is 360 degrees (360°), A half circle is 180° and a quarter of a circle is 90°.

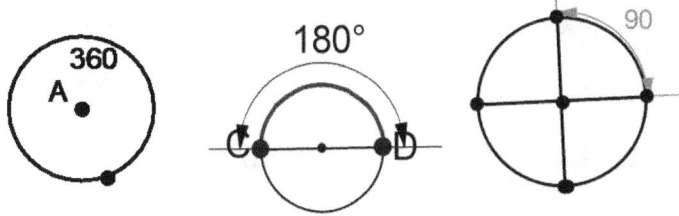

> The vertex of the angle is lined up with the origin of the Protractor which is zero degree.
> But, the protractor has two scales: the inside scale and the outside scale.
> You have to know which one to read.

o If the angle is more than 90° use the inner measure.
o If the ,measure is less than 90°, use the inner angle. It is 60
o To tell the measure of an angle, use $m \angle ABC$ which means "the measure of angle ABC?

Find the measures of these angles:

1) <PON =_____

2) <POM=_____

3) <POL=_____

4) <POK =_____

5) <KON=_____

6) <LOM=_____

7) <KOL= _____

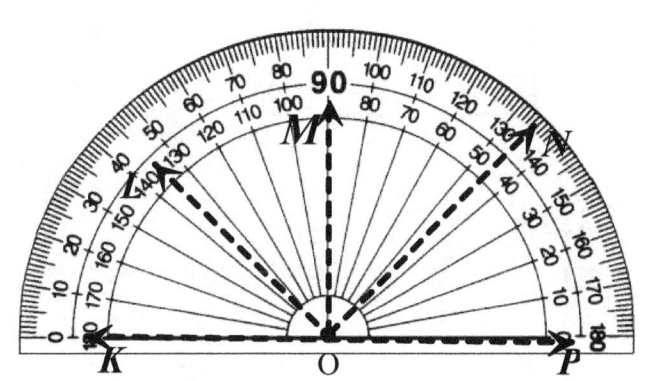

9.4) CLASSIFICATION OF ANGLES

Angles are usually classified based on their measurements.

1) **Right Angle:** Two perpendicular lines make

a right angle which is ninety degree **(90°).**

2) **Acute Angle** is less than 90°.

3) **Obtuse Angle** is greater than 90°

4) **A Straight Angle** is 180°

5) **Reflex Angle:** is an angle greater than 180

but less than 360

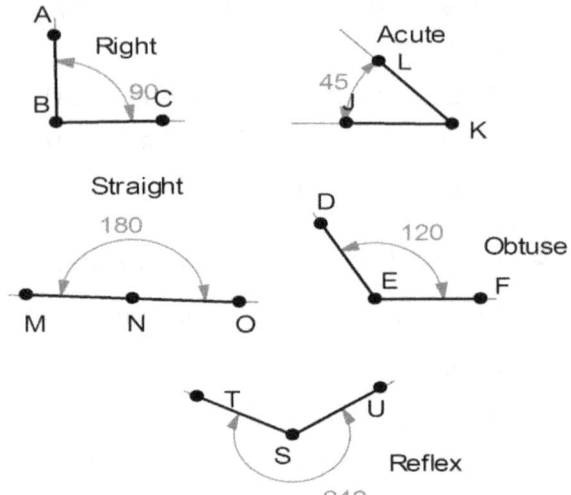

Practice: Classify the angles 1,2,3 as acute, obtuse or right. For problems 4,5 and

6, find the missing angles (x,y and z).

1)

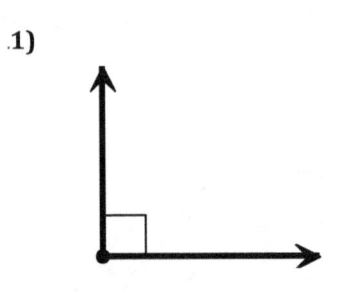

2)

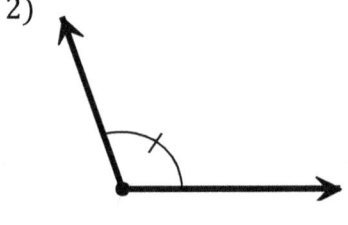

3)

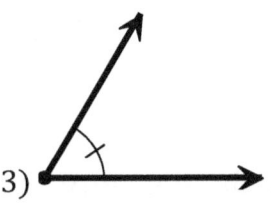

4)

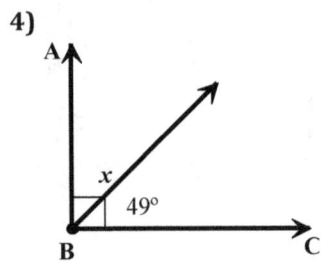

5)

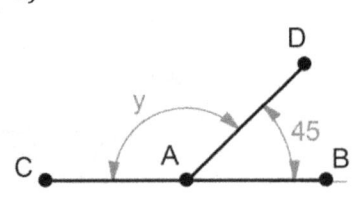

6)

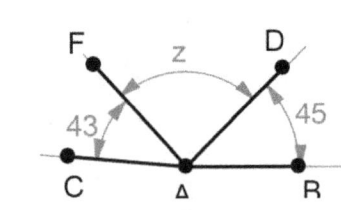

9.5) NAMING ANGLE PAIRS

Complementary Angles: Two angles that add up to 90°.

Supplementary Angles: Two angles are complementary if their sum is 180

Vertical Angles: the opposite angles 1 and 3 as well as angles 2 and 4 are vertical angles Vertical anglesa are **congruent (equal)**

AdjacentAngles: <5 and <6 are next to each other and are called adjacent angles. Could you name another pair?

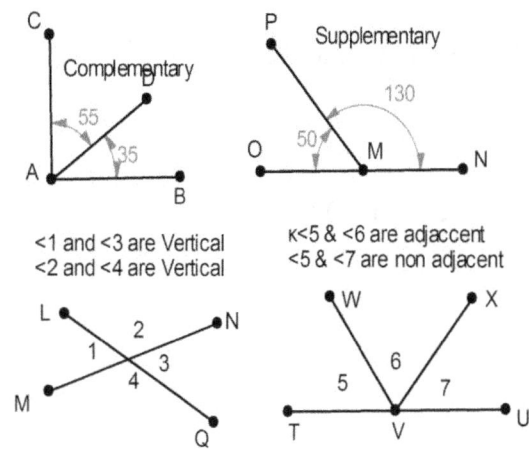

<1 and <3 are Vertical
<2 and <4 are Vertical

к<5 & <6 are adjaccent
<5 & <7 are non adjacent

Alternate Exterior Angles: <1 and <8 are alternate exterior angles and are congruent Could you name another pair?

Alternate Interior Angles: <3 and <6 are alternate interior angles and are congruent Could you name another pair?

Corresponding Angles: <2 and <6 are corrresponding angles and are congruent. Could you name another pair?

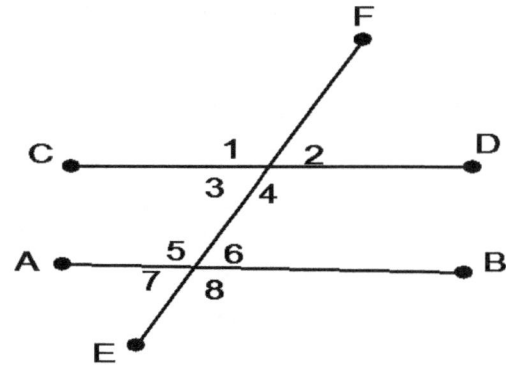

Practice: Naming Angle Pairs

In the figure below, assume line l is a transversal. Identify three pairs of :

1) Vertical Angles

2) Adjacent Angles

3) Supplementary Angles

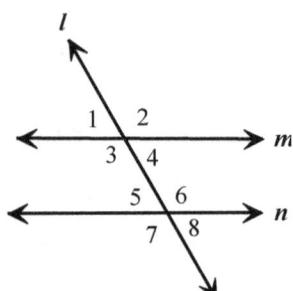

Use the figure below for problems 4 to 8

4) Name two interior alternate angles:

5) Name two exterior alternate angles

6) Name two vertical angles

7) Two corresponding angles

8) Name two supplementary angles

9) Find the measure of angle x.

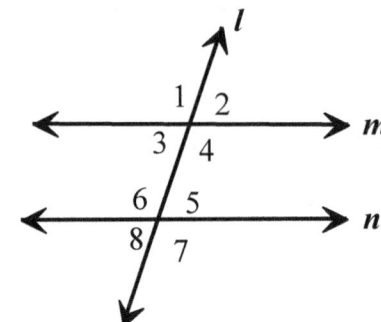

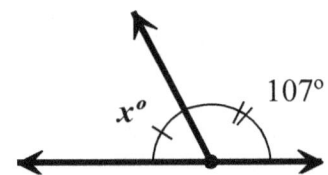

9.6) TRIANGLES

Triangles: are figures that have three sides and three angles.

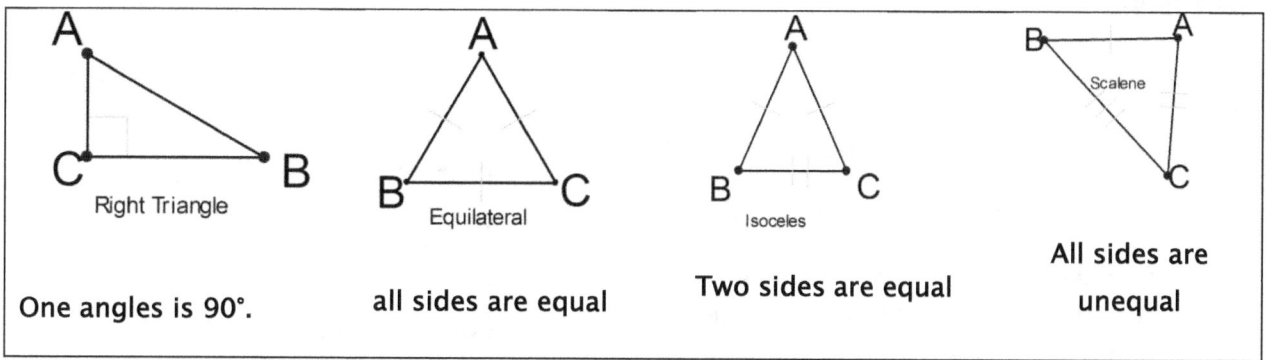

Right Triangle — One angles is 90°.

Equilateral — all sides are equal

Isoceles — Two sides are equal

Scalene — All sides are unequal

***Important: The sum of angles of any trianle is 180°.*

A) **Classify the following triangles** as equilateral, Isosceles or scalene: remember the marks show sides that are different or similar.

1)
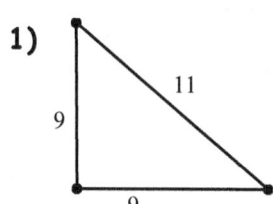

2)
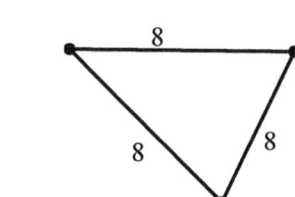

3)
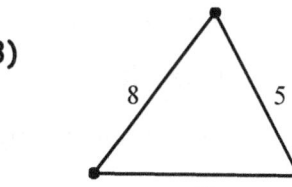

4)
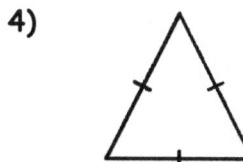

5)

6)
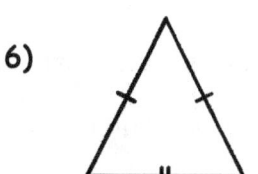

B) Find the missing angle:

7)
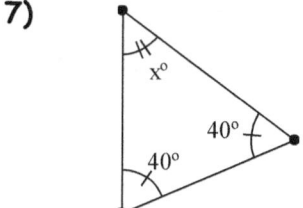

8)
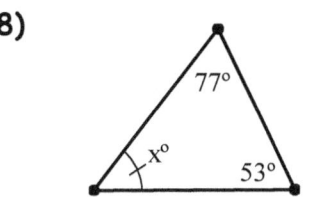

9)
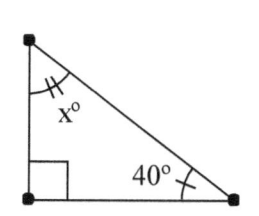

9.7) QUADRILATERALS

Quadrilaterals: are figures that have four sides: quad means four and lateral means side.

1) Squares: have four equal sides and all of its angles are also right angles.

2.) Rectangles: have each two pairs of parallel lines equal. All angles are also right angles.

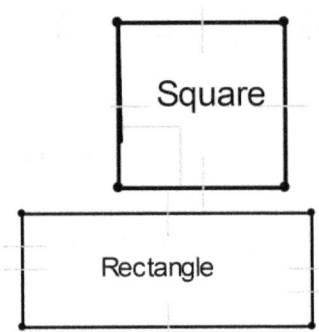

3) Trapezoid: has only one pair of parallel sides

4) Parallelogram: Each two opposite sides are equal. Opposite angles are also equal

5) Rhombus: The two pairs of opposite sides are parallel and all sides are equal.

6) Kite: the two pairs of adjacent sides are equal.

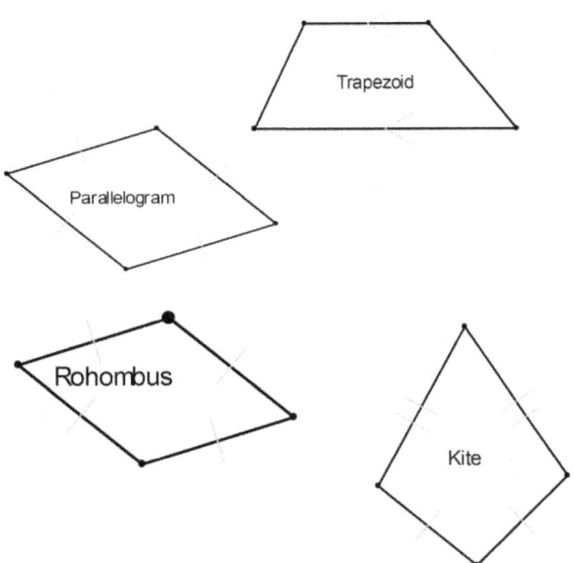

9.8) AREAS AND PERIMETERS OF POLYGONS

A **perimeter (P)** is the distance around the polygon. To find it, add the lengths of all sides.

The Area (A) of Polygons can be found in different ways (See Examples below)

Examples: Find the area and the perimeter for each of the following polygon (Assume all units are in feet).

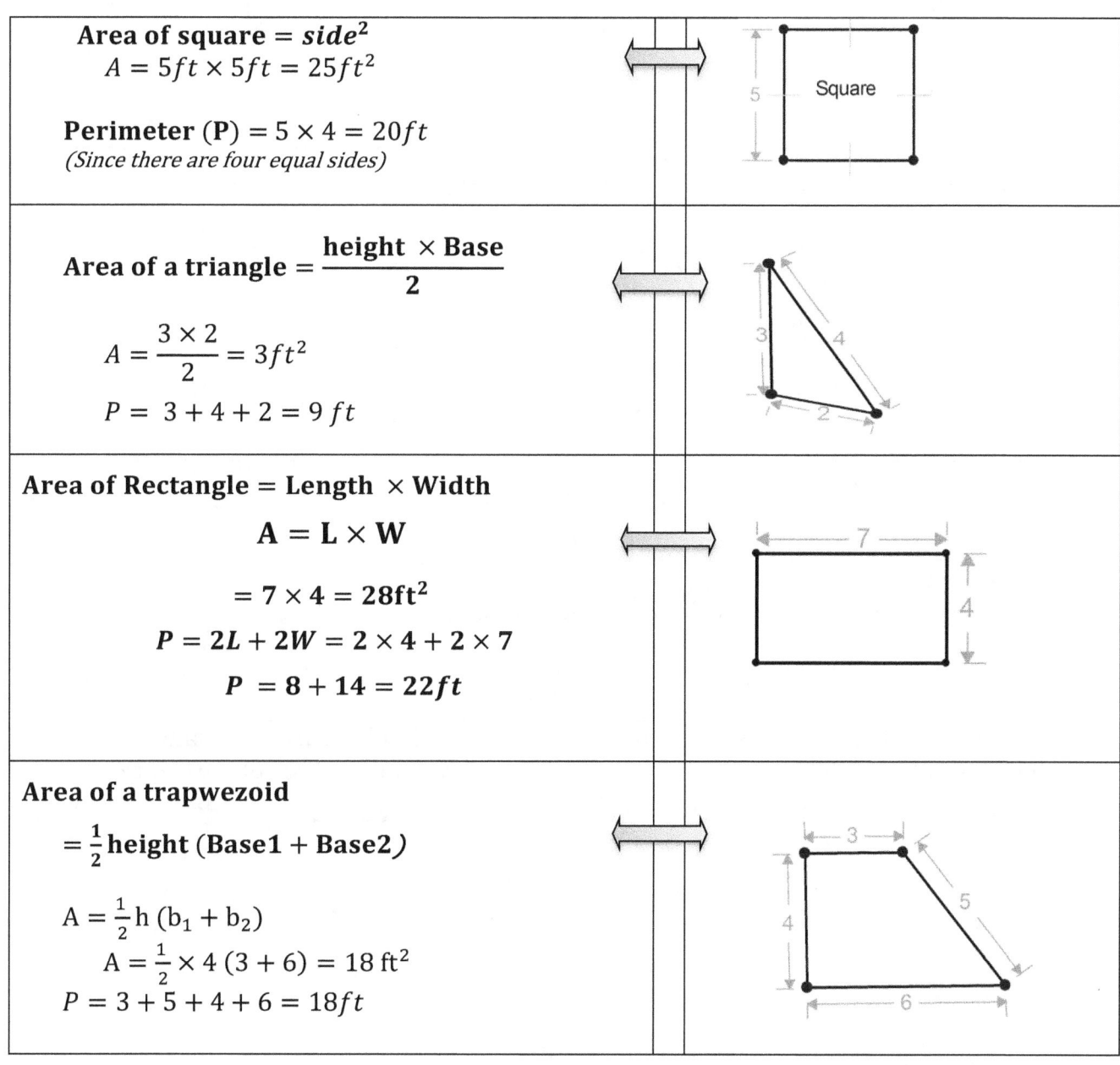

Area of square = $side^2$
$A = 5ft \times 5ft = 25ft^2$

Perimeter (P) = $5 \times 4 = 20ft$
(Since there are four equal sides)

Area of a triangle = $\dfrac{\text{height} \times \text{Base}}{2}$

$A = \dfrac{3 \times 2}{2} = 3ft^2$

$P = 3 + 4 + 2 = 9\,ft$

Area of Rectangle = Length × Width

$$A = L \times W$$

$$= 7 \times 4 = 28ft^2$$

$$P = 2L + 2W = 2 \times 4 + 2 \times 7$$

$$P = 8 + 14 = 22ft$$

Area of a trapwezoid

$= \dfrac{1}{2}$ **height (Base1 + Base2)**

$A = \dfrac{1}{2}h\,(b_1 + b_2)$
$A = \dfrac{1}{2} \times 4\,(3 + 6) = 18\ ft^2$
$P = 3 + 5 + 4 + 6 = 18ft$

Area and Perimeter Practice

Problems 1 to 9: Find the areas and the perimeters. Unless shown, assume all units are inches:

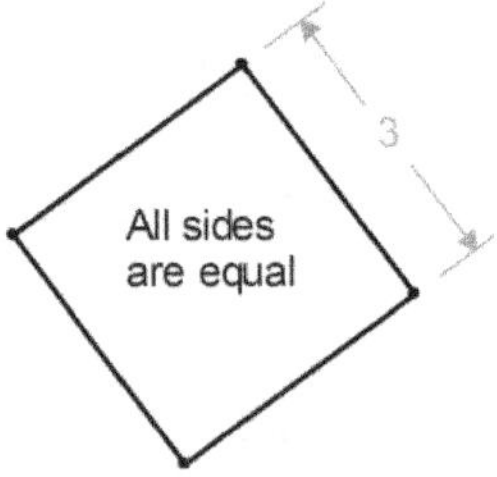

1) All sides are equal — 3

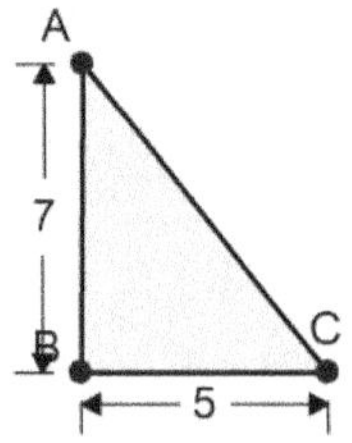

2) A, 7, B, C, 5

3) 10, 5, 4, 7

4) A, 8, D, 5, 5, B, C

5) 8 in, 8 in, 14 in

6) 6 cm, 2 cm, 3 cm, 6 cm, 8 cm

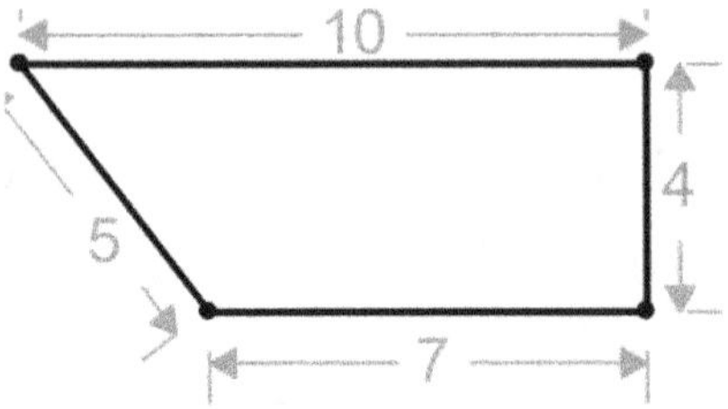

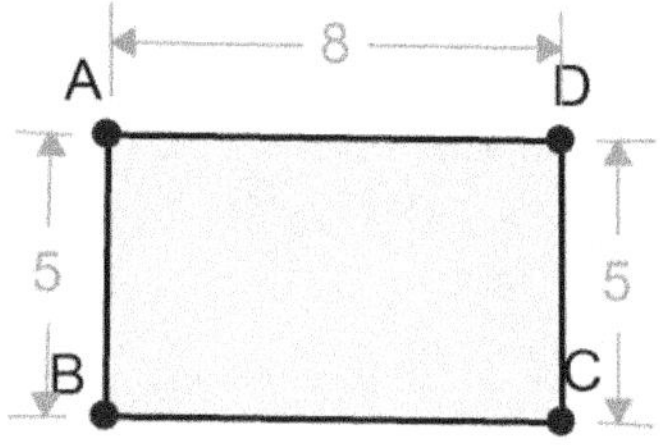

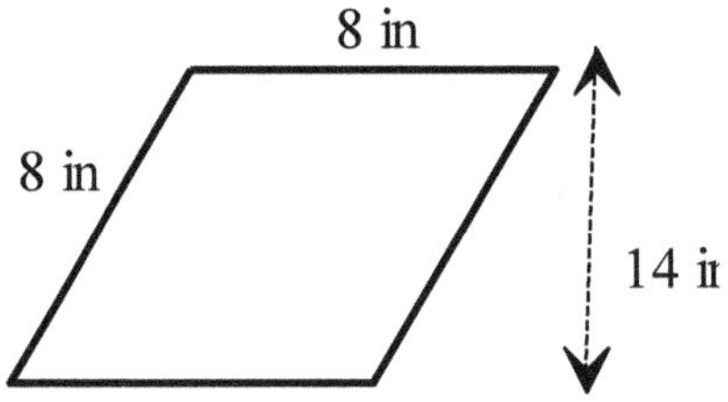

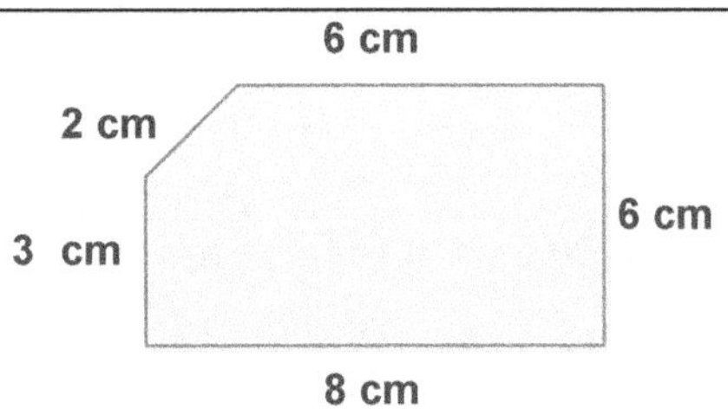

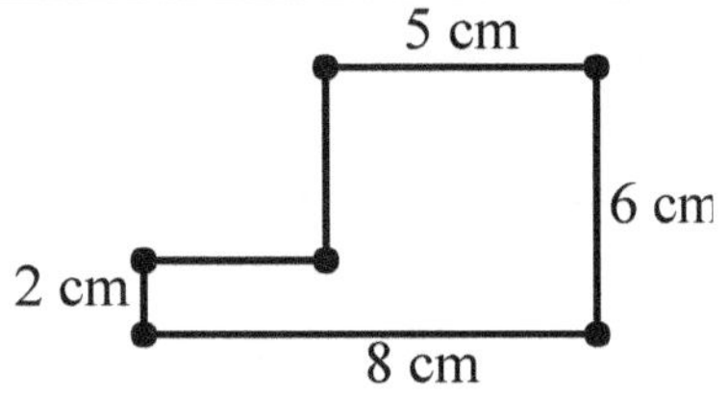

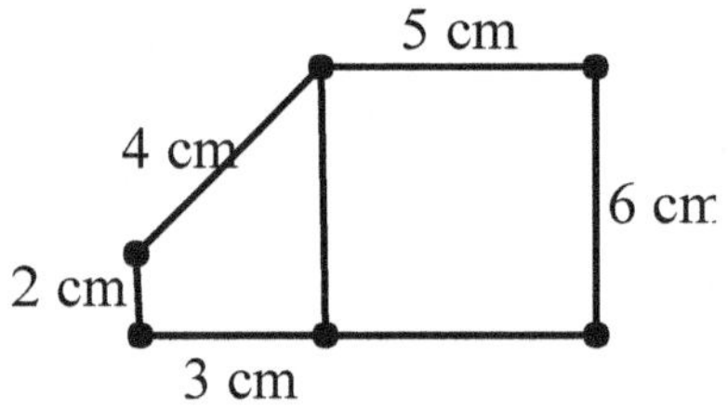

7) 5 cm, 6 cm, 2 cm, 8 cm

8) 5 cm, 4 cm, 2 cm, 6 cm, 3 cm

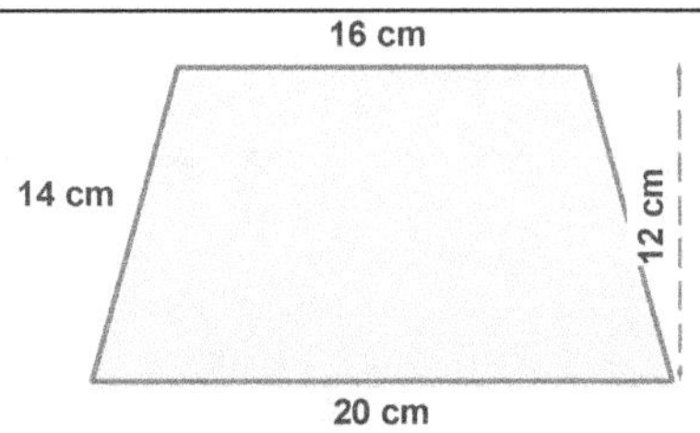

9) 16 cm, 14 cm, 12 cm, 20 cm

The premeter is 18, find the length of the missing side HK?

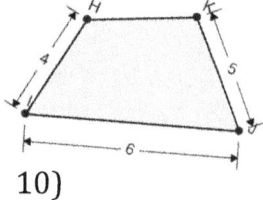

10)

(a) Find the area of triangle JHG
(b) Find the area of the trapezoid.

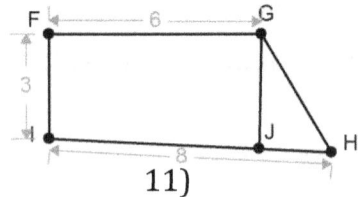

11)

9.9) THE CIRCLE

A) What is a circle? How it is named?

A circle is a shape that has all of its points the same distance from the center. Circles are named after their centers. So, for example we have here Circle A and Circle C. (Below).

B) The Diameter and The Radius

The Diameter (d) connects side to side of the circle passing through the center.

The Radius (r) is half of the diameter or the distance from the center to any points in the circle.

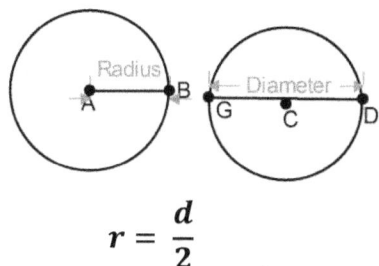

$$r = \frac{d}{2}$$

C) The Circumference and the Area:

The circumference (C) is like the perimeter: It is the distance around the circle.

$$C = 2\pi r \text{ or } C = \pi d$$

r *is the radius*

$$\pi = 3.14$$

The Area (A) of the circle is the area enclosed by the circle.

$$A = \pi r^2$$

D) Calculations: Example:

Calculate the diameter, the circumference and the area of a circle with a radius of 6 cm.

$$d = 2r = 2 \times 6 = 12 \; cm$$
$$C = \pi d = 3.14 \times 12cm = 37.68cm$$
$$A = \pi r^2 = 3.14 \times (6cm)^2 = 18.84cm^2$$

Circle Practice

Find the circumference, and the area of the following circles:

1) $d = 4cm$ $C = ?$ $A = ?$

2) $r = 5\ cm$ $C = ?$ $A = ?$

3) $d = 14\ cm$ $C = ?$ $A = ?$

4) $r = 7cm$ $C = ?$ $A = ?$

5) Name the Circles and then find the length of the diameter, the circumference and the area of each circle.

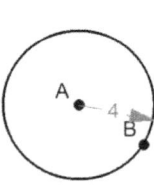 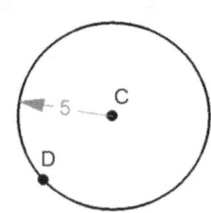 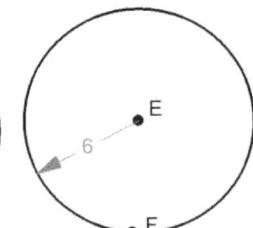

6) **True or False:** A circumference of a circle can be found by multiplying 3.14 with the diameter of the circle.

7) **True or False:** A circumference of a circle is π times its radius?

8) A pizza has a diameter of 8 inches. Find its (a) circumference and

      b) Its area.

9.10) GEOMETRY REVIEW

True or False:

1. The sum of two supplementary angles is 90°.

2. Two adjacent sides are always 180°

3. All lines must be either perpendicular or parallel

4. All angles of a square are right angles.

5. A reflex angle is more than 180°

6) Opposite sides of a parallelogram are congruent.

8) Two intersecting lines can create vertical angles.

9) Corresponding angles can be different.

10) When both bases of a trapezoid are equal, the trapezoid becomes a rectangle.

11) Find the missing angles in each figure.

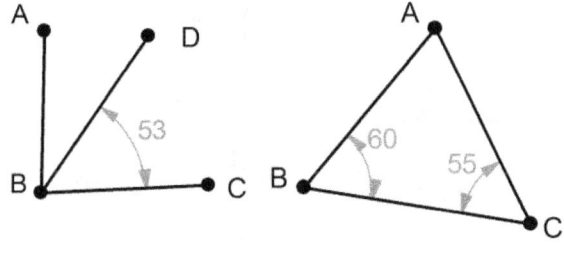

Problems 12-15: Find the perimeter and the area of the following figures:

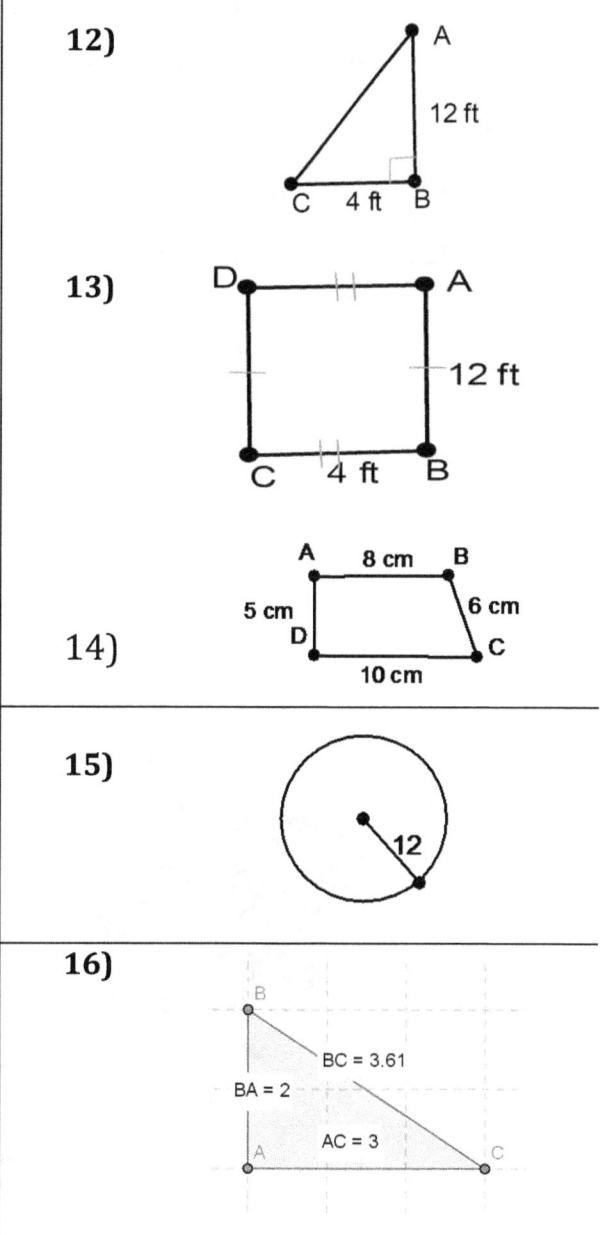

12)

13)

14)

15)

16)

UNIT 10: DATA ANALYSIS

10.1). Statistical measures: Mean, Medium, Mode & Range

10.2) The Five Number Summary: Box & Whisker

10.3) the Stem and Leaf

10.4) Different Graphs for Representing Data

10.5) How to Read Graphs

10.6) Probability of Single Events

10.1). STATISTICAL MEASURES: MEAN, MEDIUM, MODE & RANGE

Teachers usually **average** your test scores to post for you one final score. They talk about mean score. **The mean** doesn't mean a mean person etc. but just an average of, for example, your three test scores. Other common statistical measures are the **medium, the mode and the range**. Remember in general, Statistics has three main goals: collecting, summarizing, and analyzing data

Example:

Omar took seven tests and scored these grades: **calculate the mean, the medium and the mode.**

Test 1	Test2	Test3	Test 4	Test5	Test6	Test7
85	83	79	90	90	80	88

1. **The mean:** $(\bar{x})$ is the average: adding all scores and dividing by the number of tests.

$$\bar{x} = \frac{85 + 83 + 79 + 90 + 90 + 80 + 88}{7} = \frac{595}{7} = 85$$

2. **The medium:** is the middle number in the data. Write data in increasing or decreasing order.

79 80 83 **85** 88 90 90

The middle score become 85 again but it is not always equal to the mean!
Note: If the data is even you have two middle scores and you have to average the two.

3) **The mode:** is the number that shows up most. In our scores 90 shows up twice. Therefore, the mode is **90**. If two different numbers show up at the same frequency, we say the data is bimodal (means has two modes).

4) **The Range:** Is the difference between the highest score and the lowest score.

$$95 - 65 = 30$$

Practice: means, median, mode, range

1). Aisha took 5 math tests and scored: 90, 95, 80, 85 and 95.
Calculate her mean, medium, mode and range of her scores.

2) Abigail scored 90, 95, 80, 85 95, and 75 in her math 6 scores. Calculate her mean, medium, mode and range of here scores.

3) In five games in 2014 against Milwaukee, Indiana, Washington, Orlando, and Toronto, Le Bron James scored 26, 19, 29,29, 15 He played 41, 32, 36, 31, and 37 minutes respectively.

    a) What is his mean, medium, mode and range scores?

    b) What is his mean and medium minutes of play in each game?

4) Challenge: To receive an "A" in math, Mustafa has to score a mean grade of 90% in his 5 tests. He scored 85, 90, 87, 88 in his first four scores. How much should he score in the fifth score to get an "A"?

10.2) THE FIVE NUMBER SUMMARY: BOX & WHISKER

Another way to summarize data is to calculate the **Five Number Summary** which are: **Minimum** (lowest score), **Maximum** (the highest score), **Medium** (middle score), **Quartile one (Q1)** and **Quartile three (Q3). Let's use Omar's test scores again:**

Omar took seven tests and scored these grades: **calculate the five number summary.**

Test 1	Test2	Test3	Test 4	Test5	Test6	Test7
85	83	79	90	90	80	88

1. The Minimum = 79
2. **The medium or the mid number** after ranking the data **is 85:**

    79, 80, 83, **85,** 88, 90, 90

3. **Quartile 1 (Q1)** is the midpoint between the minimum and the medium. Since there are even numbers (80, and 83) we need the average of the two: **(80+83)/2= 81**

4. **Quartile 3 (Q3)** is the midpoint between the medium and the maximum. Since there are even numbers (88, and 90) we need the average of the two: **(88+90)/2= 89**

5. The Maximum = 90

> **The box and whisker plot** is simple a visual or graphical way to show the five number summary

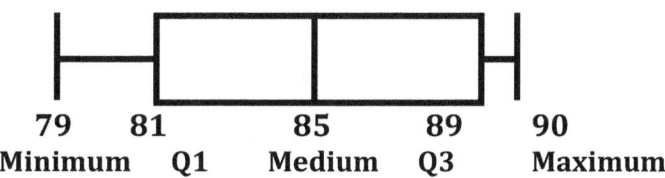

79	81	85	89	90
Minimum	**Q1**	**Medium**	**Q3**	**Maximum**

Practice the 5 number summary:

Use the following weights in pounds of a typical student weights to answer the questions below:

160, 145, 205, 150, 157, 138, 145, 210, 135, 190

a) Calculate the five number summaries of the weights.

b) Show the whisker and plot box of the data

2) Gulled visited the gym all seven days of the last week and exercised the following minutes:

60, 50, 45, 42, 55, 47, 35

a) Calculate the five number summaries.

b) Sketch the box and whisker plot.

10.3) THE STEM AND LEAF

A stem and leaf uses numbers to show the shape of a data. Let's use **an example:**

The grades of a group of students for the final math test was this:
51, 52, 54, 59, 60, 62, 62, 65, 70, 71, 71, 72, 73, ,79, 81, 82, 82, 82, 85,87, 89, 89 90, 96, 98,100

Step 1: List the stems which are the first numbers. Remember it starts with 50 to 90's. So the stem numbers are 5, 6, 7, 8, and 9

Step 2: Put the leaves: Example in the 50's the leafs are 1, 2,4 and 9.

Final Grades of Math

5	1 2 4 9
6	0 2 2 5
7	0 1 2 3 9
8	1 2 2 2 5 7 9 9
9	0 6 8 1 0 0

Key: 5 | 1   Means 51

Remember

✓ if a number is repeated more than once you should write the leaf again. Look how the 2 at 82 is repeated three times.

✓ Also write the title and key that explains what your stem and leaf represent

Practice:

The Time Spent walking to school (minutes) is shown below.

5, 10, 12, 15, 17, 30, 32, 35, 39 20, 21, 22, 25, 40

1) How many stems should your plot have?

2) How many leaves should your plot have?

3) The following are number of minutes that 25 students in certain school read every night. Draw the stem and leaf plot:

20, 21, 22, 25, 30, 31, 32, 34, 35, 40, 45, 45, 54, 55, 60, 62, 64, 65, 70, 71, 73, 75, 80, 95 97

10.4) DIFFERENT GRAPHS FOR REPRESENTING DATA

Graphs and charts make information clearer and easily understandable. Common graphs including bar charts, circle graphs, line graphs and picture graphs called also pictographs. Compare the four graphs. Which graph would you prefer?

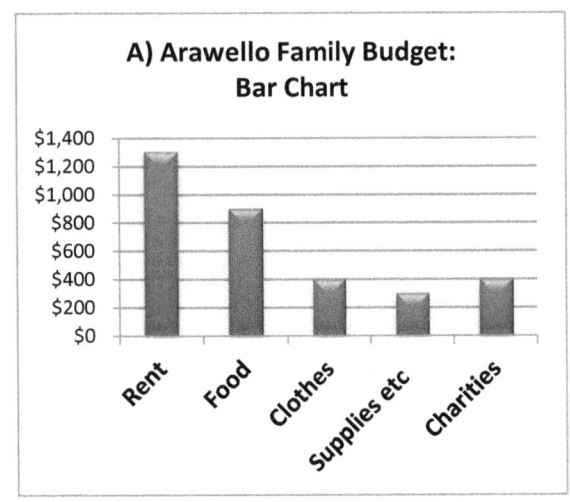

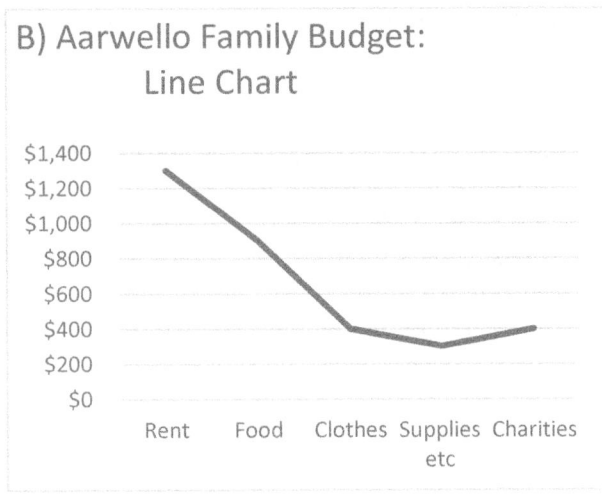

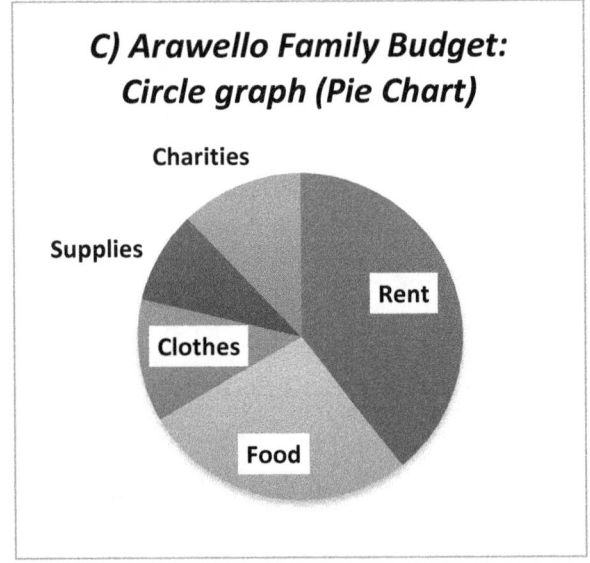

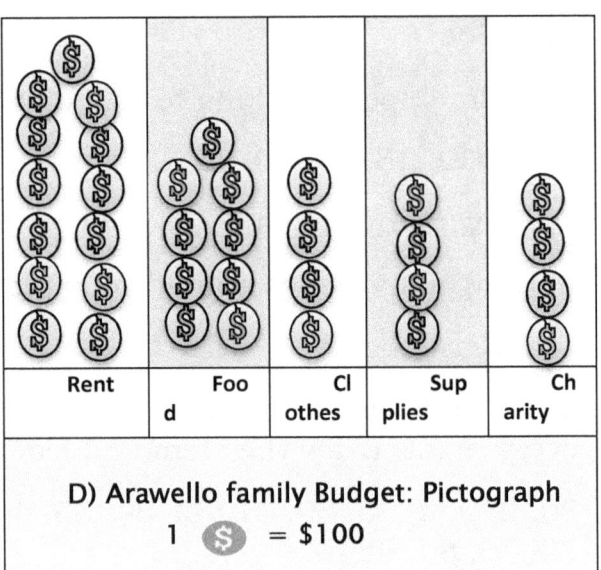

Important: A bar graph is appropriate for comparisons while line graph is good in showing changes over time. Also, Pie chart is good in data in percentages usually 6 or fewer categories. When do you think pictographs are appropriate?

10.5) HOW TO READ GRAPHS

- Look at the title of the graph
- Understand the labels (food, weight, dollars etc.)
- Estimate the numbers if needed

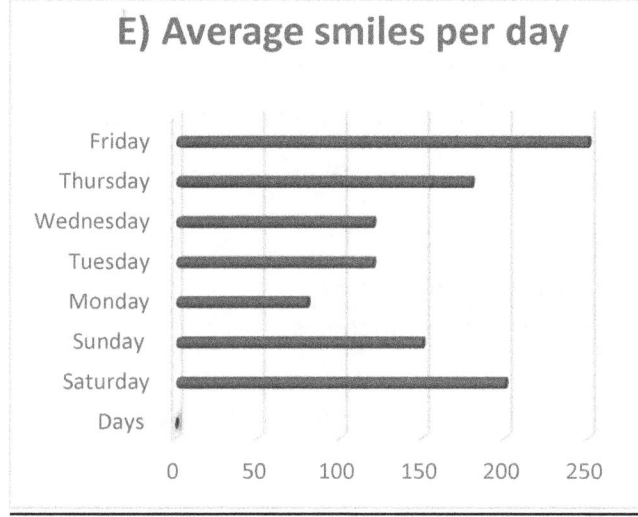

E) Average smiles per day

Use the information on graph (E) to answer questions 1 to 5.

1. Which day people smile most?

_____ _____

2. Which day people smile the least?

__ _____

3. Which two days have the same number of smiles?

_____ and _____

4. How many more smiles do people have on Friday than Saturday?

_____

5) How many times people smile on Friday, Saturday, and Sunday combined?

__ __ _____

F) Growth of Ethnic Stores in Minneapolis and Seattle

Use graph "F" on the left

6) How many more ethnic Stores were in Minneapolis in year 2 than in year 3?

_____

7) Which year the two cities had same number of ethnic stores?

_____

8) Total ethnic stores in year 4 in both cities: _____

9) How many ethnic stores were in Seattle in year 2 and year3?

Year 2_____ Year 3_____

10.6) PROBABILITY OF SINGLE EVENTS

> **An event** is the outcome that we are predicating or interested in. <u>Single Event</u> is just one event like throwing a coin or die once

> ➤ **The probability of an event** is the ratio of the number of favorable outcomes (called also successes) of an event over the total number of possible outcomes. Favorable outcome means the number of times that the event you are testing will happen.

$$➤ \; P = \frac{\text{Number of favorable outcomes}}{\text{Total number of Possible Outcomes}}$$

> ➤ Also, the probability of any event is always a fraction or a decimal between 0 and 1.
> $$0 \le p \le 1$$

> ➤ A Probability of 1 means we are 100% sure. A zero probability event is an impossible event.

Example 1: A coin is flipped in the air. What is the likelihood that it will land on its head?

A coin has 2 possible outcomes: Head or Tail

Here there is only one head. The number of successes is 1

$$P = \frac{1}{2}$$

Example 2: A die is tossed once. What is the probability of getting a number greater than 2?

A die has six possible outcomes since it has 6 sides.

All four numbers (3, 4, 5, and 6 are more than 2).

⇨The number of successes= 4

$$p = \frac{4}{6} = \frac{2}{3}$$

Use the data for problems 1 to 10: a bag contains 6 different numbers: 2, 4, 6, 8, 10, and 20. If person picks just one number:

1. What is the probability that the number is ten?

2. What is the probability that the number is less than 6?

3. What is the probability that the number is less than 2?

4. What is the probability that it is more than 20?

5. What is the probability that it is a number more than 4?

6. What is the probability that the number is 4 or more?

7. What is the probability that it is greater than 10?

8. What is the probability that that number is divisible by 5?

9. What is the probability that the number is divisible by 10?

10.   What is the probability that the number is 25?

Use this data for problems11–13: Find the probability that the spinner:

11) Will land on black

12) Will land on white?

13) Will land on yellow or black?

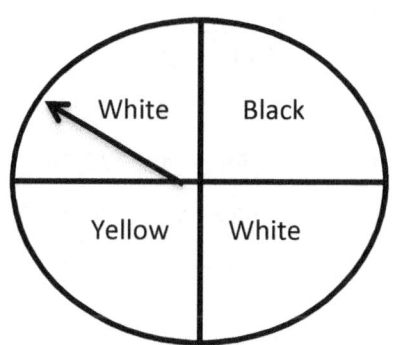

SELECTED ANSWERS

UNIT 1: THE FOUNDATION

1.4) Order of Operations: **1) 20   3) 14   5) 1   9) 2**

1.5) Review of Primes and Composite?   3) Composite   9) prime   13) prime   17) composite

1.6) Prime Factorization : 1) 2^3 3) $2^3 \times 7$ 5) 3×13 7) $2^2 \times 5^2$

1.7) Factoring: 1) 1, 2, 4, 8, 16,   5) 1, 2, 3, 4, 6, 7, 12, 14, 21, 28, 42, 84   7) 1, 2, 4, 8, 17, 34, 68, 136

1.8) Review of Perfect Squares $0^2 = 0$ $12^2 = 144$ $15^2 = 225$ $19^2 = 361$

1.9) Square Roots: $\sqrt{169} = 13$ $\sqrt{196} = 14$ $\sqrt{324} = 18$

UNIT 2: FRACTIONS REVIEW:

2.1) Greatest Common Factors (GCF): (A**) 3)**28   **5)** 22   **7)** 144   (B) **3)**7   **5)** 2   **7)** 8

2.2) Find The Least Common Multiple (LCM): (A**) 3)**7   **5)** 11   **7)** 1   (B) **3)**18   **5)** 56 **7)** 80

2.3) Review Of Fractions: Improper & Mixed 1) $1\frac{3}{4}$ 3) $5\frac{2}{3}$ 5) $3\frac{3}{8}$ 7) $8\frac{1}{2}$ 17) $\frac{16}{13}$

2.4) Simplify Fractions : 1) $\frac{1}{4}$ 3) $\frac{1}{2}$ 5) $\frac{1}{3}$ 17) $7\frac{3}{5}$ 19) $5\frac{2}{3}$ 27) $13\frac{3}{5}$

2.5) Add/Subtract Fractions: (A) 1) $\frac{3}{5}$ 3) $\frac{1}{2}$ 5) $\frac{1}{2}$ (B) 1) $1\frac{3}{8}$ 3) $\frac{13}{20}$ 5) $\frac{1}{2}$

2.6) Multiply Fractions & Mixed Numbers: 1) $\frac{2}{25}$ 3) $\frac{3}{8}$ 5) $\frac{1}{15}$ 7) $\frac{1}{12}$

2.7) Divide Fractions/Mixed Numbers: 1) $\frac{2}{25}$ 3) $1\frac{1}{3}$ 5) $\frac{1}{5}$ 7) 1

UNIT 3: THE DECIMALS

3.1) How to Add or Subtract Decimals: 1) 5.5   3) 3.64   5) 96.54   7)10.85

3.2) Add or Subtract Decimals: 1) 14.5   3) 5.66   5) 99.54   7)11.15   9) 99.54   11) 3.74

3.3) Multiplication of Decimals (3) 120.345,   1216,   6761.488,   4.3709

3.4) Division of Decimals: 3.4) A) 1) 2.9 3) 0.045 5) 12.7   13) 0.65 (B) 1) 290 3) 20 5) 50 13) 50

3.5) Decimal Languages: 1) R   3) R   5) NR   7) NR   9) NR

3.6) Change Terminating Decimals into Fractions: 1) $\frac{3}{10}$ 3) $\frac{51}{50}$ 5) $\frac{13}{20}$ 3) $7\frac{1}{20}$ 3) $9\frac{1}{2}$

3.7) Change Decimals into percent: 1) 130%   3) 102%   5) 650%   7)705%   9) 950%   11) 17.4%

UNIT 4: RATIOS, RATES AND PROPORTIONS

4.2) Ratios Problems: 1a) $\frac{4}{5}$ 3) $\frac{7}{13}$ 5) $\frac{412}{147}$ 7) $\frac{5}{14}$

4.3) Equivalent Ratio Tables: 1) 8, 24, 4   3) 6, 30, 24

4.4) Unit Rates & Unit Costs 1) 67 words/ minute   3) 40 miles/hour   5) $2.7/ gallons   7) $35

4.5) Proportions: Equivalent ratios: 1) 16   3) 12   5) 8   7) 2   9) 5

4.6) Are Two Ratios Equal? 1) Yes   3) No.   5) No   7) yes   9) No   13) d

4.7) Solve Proportions: 1) 12   3) 36   5) 9   7) 2   9) 12   11) b

UNIT 5: PERCENT, FRACTIONS AND DECIMALS RELATIONSHIPS

5.1) what is Percent?   1) 25%   3) 10%   5) Playing & reading 7) 75%

5.2) Percent Concept Graphics: 1) 34%   3) 41%   5) 37.5%   7) 66.7%   9) 25%   11) 100%

5.3) Change Fractions into Percent 1) 75%   3) 25%   5) 20%   7) 83.3%   9) 8.3%   11) 41.7%

5.4) Change Mixed Numbers into Percent: **A)** 1) 350%   3) 725%   5) 920%   7) 720%   9) 1125%

5.4) Change Percent into Fractions: 1) $\frac{13}{100}$ 3) $\frac{17}{100}$ 5) 2   7) $\frac{41}{100}$ 9) $\frac{11}{20}$

5.5) Change Mixed Number Percent into Fractions) 1) 2.25   3) 5.25   5) 1.6   7) 8.25   9) 12.40

5.6) Change Percent into Decimals: A) 1) .13   3) .17   5) 0.013   7) 2.0   9) 4.5.

5.7) Change Decimal Percent into Fractions   1) $\frac{13}{1000}$ 3) $\frac{17}{1000}$ 5) $\frac{1}{50}$ 7) $\frac{11}{5000}$

5.8) Mixed Review: Fractions, Decimals, Percent 2) 0.2, 20% 3) 7/10, 70% 5) 0.75, 75% 7) 3/200, 0.015

UNIT 6: PERCENT PROBLEMS

6.1) Percent: Finding the Part: 1) 4 3) 75 5) 41 7) 168 9) 0.5

6.2) Finding the Percent : 1) 4 3) 25 5) 4.76 7) 2.5 9) 2

6.3) Finding the base: 1) 300 3) 300 5) 500 7) 1000 9) 470

6.4) Find the Percent Change: 1) 20% 3) 16% 5) 75% 7) 7.9%

6.5) Percent: Mixed Practice: 1) 10 3) 4% 5) 12.5% 7) 40% 9) 80%

UNIT 7: CONSUMER MATH: TAXES, DISCOUNTS AND INTERESTS 65

7.1) Calculate the Tax/ Other increases: 1) 480 3) 2750 5) 336   6) 14.16

7.2) Calculate the Discounts/ Other Decreases: 1) 27   3) 507   5) 11% discount

7.3) The Simple Interest : 1) $208.8 3) $907.2 5) $280 7) 25 years

7.4) Compound Interest : 1) $1098.35   3) $11902.38   5) $9231.7

UNIT 8: MEASUREMENTS

8.2) US measures: Units of Length: 1) 180 in 3) 24 yd. 5) 18,480 ft. 7)3 yd 1ft. 9) 1.67 mi

8.3) Metric System: Units of Length: 1) 1500 cm. 3) 2200 mm 5) 36 km 7) 250 mm 9) 8.8 km

8.4) Conversion Metric & U.S Length: 1) 196.9 in 3) 15.7 cm 5) 396.8 mi 7) 9.9 in 9)985 in

8.5) Units of Weight: 3) 4 kg 5) 226796 g 7) 9080 g 9) 5.3 lb

8.6) Units of Capacity or Volume: 3) 4 *l* 5) 6 *l* 7) 1892 ml 9) 16 ml.

8.7) Add and Subtract Mixed Units: 1a) 17ft 1 in   1b) 2 yd. 1c) 13m. 10cm. 3a) 8g, 300mg. 3b) 5c 5fl.oz. 3c) 1ft 4 in   5a) 5ft 4in 5b) 3 yd 2ft 5c) 4m 70cm

UNIT 9: GEOMETRY

9.2) How to Name Angles: 1) <AOE or <EOA   3) <DOC or COD 5) <BAC, <CAB or <A 7) <IGH, <HGI or <G

9.3) How to Measure Angles: 1) 45° 3) 135°   5) 135°   7) 45°

9.4) Classification of Angles 1) **right 2) Obtuse**   3) **acute**   4) **x=41**   5) **y=135**   6) **z=92**

9.5) Naming Angle Pairs : 1) 2&3, 1&4, 5&8   3) 1&2, 3&4, 5&6 5) 1&7, 2&8.   7) <1 &<4.   9)73°

9.6) Triangles   : 1) isosceles   3) scalene   5) scalene 7) 100° 9) 50°

9.8) Areas and Perimeters of Polygons: 1) 9 in^2; 12 in 3) 34 in^2;26 in 5) 128 in^2;32 in 7) 36 cm^2;28 cm 9)216cm^2; 64cm   11) Area of triangle= 3 in^2; Area of Trapezoid= 21 in^2

9.9) The Circle: 1) C= 12.56 cm; A = 12.56 cm^2 3) C= 43.96 cm; 153.86 cm^2 5) Circle A; d=8, C = 25.12 cm , A = 50.24cm^2 7) False

9.10) Geometry Review : 1) False 3) False 5) False   7) True 9) True 11) 24 ft^2 13) 45cm^2 15) 3

UNIT 10: DATA ANALYSIS   93

10.1). Statistical measures: 1) mean = 89, med =90, mode=95, Range= 15   4) 100

10.2) the 5 Number Summary: 1) Minimum: 135, Quartile Q1: 145, Median: 153.5, Q3: 175, Maximum: 210

10.3) Stem and Leaf   1) 4   2) 14

10.5) How to Read Graphs 1) Fridays   3) Wednesdays and Tuesdays 5) 600 7) year 1 9) 30, 38

10.6) Probability of Single Events: 1) $\frac{1}{6}$ 3) 0   5) $\frac{2}{3}$ 7) $\frac{1}{6}$ 9) $\frac{1}{3}$ 11) $\frac{1}{4}$ 13) $\frac{1}{2}$

You did IT!

Congratulations!

www.ingramcontent.com/pod-product-compliance
Lightning Source LLC
Chambersburg PA
CBHW080620190526
45169CB00009B/3244